SpringerBriefs in Computer Science

SpringerBriefs present concise summaries of cutting-edge research and practical applications across a wide spectrum of fields. Featuring compact volumes of 50 to 125 pages, the series covers a range of content from professional to academic.

Typical topics might include:

- A timely report of state-of-the art analytical techniques
- A bridge between new research results, as published in journal articles, and a contextual literature review
- A snapshot of a hot or emerging topic
- An in-depth case study or clinical example
- A presentation of core concepts that students must understand in order to make independent contributions

Briefs allow authors to present their ideas and readers to absorb them with minimal time investment. Briefs will be published as part of Springer's eBook collection, with millions of users worldwide. In addition, Briefs will be available for individual print and electronic purchase. Briefs are characterized by fast, global electronic dissemination, standard publishing contracts, easy-to-use manuscript preparation and formatting guidelines, and expedited production schedules. We aim for publication 8–12 weeks after acceptance. Both solicited and unsolicited manuscripts are considered for publication in this series.

**Indexing: This series is indexed in Scopus, Ei-Compendex, and zbMATH **

Meng Jiang • Bill Yuchen Lin • Shuohang Wang •
Yichong Xu • Wenhao Yu • Chenguang Zhu

Knowledge-augmented Methods for Natural Language Processing

 Springer

Meng Jiang 🆔
Department of Computer Science
and Engineering
University of Notre Dame
Notre Dame, IN, USA

Shuohang Wang
Microsoft
Redmond, WA, USA

Wenhao Yu
Department of Computer Science
and Engineering
University of Notre Dame
Notre Dame, IN, USA

Bill Yuchen Lin
Allen Institute for AI
Seattle, WA, USA

Yichong Xu
Microsoft
Redmond, WA, USA

Chenguang Zhu 🆔
Microsoft
Redmond, WA, USA

ISSN 2191-5768   ISSN 2191-5776   (electronic)
SpringerBriefs in Computer Science
ISBN 978-981-97-0749-2   ISBN 978-981-97-0747-8   (eBook)
https://doi.org/10.1007/978-981-97-0747-8

This Springer imprint is published by the registered company Springer Nature Singapore Pte Ltd.
The registered company address is: 152 Beach Road, #21-01/04 Gateway East, Singapore 189721,
Singapore

Paper in this product is recyclable.

Preface

Ever since language was invented and used, humans have leveraged this powerful tool to pass down experience over generations. These kinds of knowledge summarize precious findings and inspiring ideas, which constitute the human wisdom.

Fast forward to the current days, we have for the first time witnessed computerized models, commonly known as Large Language Models (LLMs), that can understand and produce human language to a degree that is widely acknowledged and adopted. One may raise the question: is natural language processing (NLP) solved?

Many evidences have pointed out that while LLMs are excellent in language-related capabilities and manifest sparks of intelligence, they still struggle to understand deep logic, human experience and incessantly incoming new information. Therefore, it is vital to infuse human knowledge into these LLMs to further reduce the gap between its intelligence level with humans' level.

Since 3 years ago, way before the upsurge of LLMs, I have started the research of knowledge-augmented NLP together with the other authors of the book. We have found that various forms of knowledge, e.g., free-form text, tables, knowledge graph, dictionary, could effectively enhance the performance of language models in many NLP tasks. These findings become even more important after the advent of LLMs, since users put more faith into their capabilities while LLMs alone without accurate and fresh external knowledge can often bring disappointment.

In view of this, we have published papers, hosted workshops and tutorials, to address the urgency and methods of knowledge-augmented NLP. This book summarizes both our findings and exciting progress in this field. We hope that this book can bring valuable insights and encourage more people to conduct related research and push forward the boundary of NLP models.

This book has six chapters, including a variety of topics such as knowledge augmentation in natural language understanding and generation, knowledge sources and commonsense knowledge integration. And they are written by:

Chapter 1 (Introduction to Knowledge-Augmented NLP): Chenguang Zhu

Chapter 2 (Knowledge Sources): Shuohang Wang

Chapter 3 (Knowledge-Augmented Methods for Natural Language Understanding): Yichong Xu

Chapter 4 (Knowledge-Augmented Methods for Natural Language Generation): Wenhao Yu

Chapter 5 (Augmenting NLP Models with Commonsense Knowledge): Bill Yuchen Lin

Chapter 6 (Summary and Future Directions): Meng Jiang

Last but not least, we want to sincerely thank Springer Nature editors Sudha Ramachandran and Nick Zhu for their great support in writing this book.

Issaquah, WA, USA   Chenguang Zhu
November 2023

Contents

Chapter 1
Introduction to Knowledge-augmented NLP

Abstract There has been tremendous progress on the research of Natural Language Processing since more than half a century ago. While the latest development of Large Language Models (LLMs) has brought unprecedented enthusiasm about achieving human-level language understanding and generation, there still exist considerable limitations of these language models, such as the lack of world knowledge, explainability and generalization. To solve these issues, it is important to augment NLP models with external knowledge sources, including both unstructured knowledge, e.g., free-form text, and structured knowledge, e.g., knowledge graphs. The integration of these knowledge sources consists of three steps: (1) Grounding language into related knowledge; (2) Representing knowledge; and (3) Fusing knowledge representation into language models.

Keywords Natural language processing · External knowledge · Structured knowledge · Unstructured knowledge · Grounding · Representation · Fusing

1.1 A Brief History of NLP

Natural language processing (NLP) is the field of computer science and artificial intelligence that aims to enable machines to understand, generate, and interact with natural languages. The history of NLP can be dated back to the early days of computing and linguistics, and has evolved through different paradigms, challenges, and applications over the decades.

The first era of NLP is dominated by the rule-based or symbolic approach, which relied on manually crafted grammars, dictionaries, and logic to represent and manipulate natural language. One of the earliest and most influential works in this era was the Automatic Translation Project, initiated by Warren Weaver in 1949, which proposed to use mathematical and logical methods to translate texts between languages. This prompted the result in more NLP areas. However, a report [1] in 1966 by the Automatic Language Processing Advisory Committee (ALPAC) concluded that machine translation was slower, less accurate, and more expensive

© The Author(s), under exclusive license to Springer Nature Singapore Pte Ltd. 2024
M. Jiang et al., *Knowledge-augmented Methods for Natural Language Processing*,
SpringerBriefs in Computer Science, https://doi.org/10.1007/978-981-97-0747-8_1

than human translation, and recommended to reduce the funding and focus on basic research instead. The ALPAC report marked the end of the first era of NLP.

The second era of NLP, which emerged in the 1980s and 1990s, was characterized by the statistical or data-driven approach, which leveraged the availability of large corpora of texts and speech, and the advances in machine learning and probabilistic models, to learn patterns and rules from data, rather than relying on human experts. This approach enabled NLP to handle more realistic and diverse language phenomena, such as word sense disambiguation, part-of-speech tagging, parsing, named entity recognition, sentiment analysis, and machine translation. The statistical approach also improved the performance and efficiency of NLP systems, and facilitated the evaluation and comparison of different methods and systems using quantitative metrics and benchmarks.

The third and current era of NLP, which started in the late 2000s and continues to the present day, is driven by the neural or deep learning approach, which exploits the availability of massive amounts of data, the advances in computational power and hardware, and the development of novel architectures and algorithms, to build end-to-end models that can learn complex and high-level representations and functions from raw inputs, such as words, characters, or speech signals. The neural approach has achieved remarkable results and breakthroughs in various NLP tasks as well as enabling the development of general-purpose and pre-trained models, such as recurrent neural networks, convolutional neural networks, attention mechanisms, transformers, large language models (LLMs) such as BERT [2], T5 [3] and GPT [4–6]. These models can capture rich contextual information from language, and can be finetuned or adapted to specific tasks and domains. Most prominently, large language models have captured most attention and been deemed promising for various intelligence tasks beyond the traditional NLP domain, such as solving math problems, writing code, complex task planning, etc.

1.2   What Large Language Model is Not Good At

Large language models represent the latest and highest achievement in NLP. However, there are still inherent limitations that prevent LLMs from fully understanding and generating natural language texts. Some of these limitations are:

Lack of common sense and world knowledge. LLMs rely on the statistical patterns and co-occurrences of words and phrases in the text data, but they do not have access to the factual, conceptual, and causal knowledge that humans frequently use to reason and communicate. This complete reliance on statistical patterns extracted from limited data with a cutoff date is bound to lead to errors or inconsistencies when LLMs generate or interpret texts that require knowledge not represented or present in the training corpus.

Lack of explainability and transparency. LLMs are often regarded as black-box models, as it is difficult to understand how they make decisions or what features they use to represent the meaning of texts. This can pose challenges for debugging, evaluating, and trusting LLMs, especially in sensitive or critical applications, such as medical diagnosis, legal analysis, or ethical reasoning. Moreover, LLMs may inherit or amplify the biases, errors, or noise in its training data, which can affect their fairness, accuracy, and reliability.

Lack of generalization and adaptation. LLMs are trained on specific text data, which may not cover all the possible scenarios, domains, or tasks that they encounter in real-world applications. For example, LLMs may not be able to handle texts that are in different languages, styles, or formats, or that involve new concepts, entities, or relations. Moreover, LLMs may not be able to learn from new or dynamic information, such as user feedback, online updates, or contextual cues, which can affect their relevance, timeliness, and personalization.

These limitations suggest that LLMs need to be augmented with external knowledge sources that can provide them with more information, structure, and guidance.

1.3 Common Knowledge Sources

Given the goal of augmenting LLMs with necessary evidence, we define knowledge in this context as any information that is absent from task input but helpful for generating the output. Thus, there is a broad spectrum of knowledge sources that can be leveraged to provide rich and up-to-date information about the world and domains of interest. In general, they can be categorized into the two following categories.

Unstructured Knowledge. This mainly refers to free-form text. Typical sources include webpages and books/papers, such as the Wikipedia and BookCorpus data used in pre-training BERT. For private domains, business documents are another important source. Furthermore, given the vast amount of knowledge grasped by LLMs, they can in turn become the source of knowledge. For instance, one can prompt a GPT model with the task input to obtain related knowledge in its textual output.

Structured Knowledge. This mainly refers to information organized with a pre-defined data model. Typical sources include knowledge graphs, databases, knowledge bases, ontologies and dictionaries. The inherent structure often brings additional information beyond text, such as hierarchy, properties and relations.

The benefits of these external knowledge sources are threefold. First, they can be stored separately from the LLM, and efficiently retrieved using various algorithms. Therefore, it can drastically reduce the requirement of a large parameter space in LLM, which has been equipped with the onerous task of implicitly memorizing all knowledge in the vast training data. Secondly, humans can directly comprehend and update these knowledge sources as they are readily readable and understandable. This is contrary to the black-box style of LLMs, which is one of the most outstanding problems criticized by the community. Thirdly, by learning and referring to external knowledge, the decision of LLMs become more explainable and transparent. In case of an undesired output, one can quickly tell whether the knowledge needs revision, or the correct knowledge is not retrieved, or the LLM makes a wrong inference given the correctly retrieved knowledge.

1.4 How to Integrate Knowledge into Language Model

Just as there exist various forms of knowledge, there are different ways to integrate knowledge into language model, depending on the format of knowledge, requirement of computation complexity, language model characteristics. In general, there are three main steps in knowledge integration: (1) Grounding language into related knowledge, (2) Representing knowledge, and (3) Fusing knowledge representation into language model.

Grounding Language into Related Knowledge As the amount of existing knowledge is huge, it is neither practical nor necessary to resort to the entirety of knowledge for each task input. Thus, the first step in knowledge integration is to ground natural language input into the related knowledge. For instance, given an input sentence, the grounding finds in a knowledge graph the entities and relations related to the input. This step can be accomplished by simple rule-based methods such as pattern matching and SQL statement creation, and also by more sophisticated methods such as named entity recognition, information retrieval and embedding-based search.

Representing Knowledge The second step in knowledge integration is to represent and encode the grounded knowledge to be consumed in the language model. This process should (i) retain the semantic and structural information of the knowledge source and, (ii) provide a compact and expressive representation of the knowledge that can be easily integrated with the language representation. In general, depending on the level of abstraction and generalization of the knowledge, there are two types of representation methods:

1. Symbolic methods, which use discrete and logical symbols, such as text identifiers, predicates, or rules, to represent knowledge entities and relations. These methods are simple, precise and interpretable by human, but they need to be later converted into continuous and distributed representations of the language model. Furthermore, symbolic representation may not capture complex structural relation in the knowledge source.
2. Vector-based methods, which use continuous and distributed vectors, such as embeddings, to represent knowledge. The representation could be contextual word embedding for unstructured text knowledge, or graph embeddings for structured knowledge graphs. Vector-based methods are compatible and efficient with the language model, but they may lose some of the interpretability and precision of the knowledge.

Fusing Knowledge Representation into Language Model The third step in knowledge integration is to fuse the knowledge representation into the computation of language model. The fusing method highly depends on the format of knowledge representation in the second step. For symbolic representation, the knowledge could be directly concatenated with the language input. It saves the need for any modification to the language model structure, thus applicable to any language model accepting text input. However, the added knowledge text may interfere with the comprehension of the original input. For vector-based representation, one can adopt the attention mechanism to weight and combine the embeddings of knowledge and language input. These attention-based methods are flexible and powerful, but they require modification to the language model structure and may introduce inconsistency in embedding distribution, at least in the beginning of knowledge fusing. Other than attention-based methods, one can employ fusion operations, such as gating, interpolation, or aggregation, to integrate knowledge representation into language models. One example is to leverage knowledge embeddings to modulate or augment the hidden states or outputs of the language model. These methods are adaptive and robust, but they introduce additional parameters or hyperparameters which may complicate the learning process.

In real applications, the decision to choose knowledge source, representation and fusing methods is dependent on the availability of knowledge, data distribution, language model structure and computational resource constraints. And it is not uncommon that multiple methods are applied to maximize the performance gain from integrating external knowledge.

References

1. Hutchins, J.: Alpac: the (in) famous report. Read. Mach. Transl. **14**, 131–135 (2003)
2. Devlin, J., Chang, M.-W., Lee, K., Toutanova, K.: Bert: Pre-training of deep bidirectional transformers for language understanding. Preprint (2018). arXiv:1810.04805
3. Raffel, C., Shazeer, N., Roberts, A., Lee, K., Narang, S., Matena, M., Zhou, Y., Li, W., Liu, P.J.: Exploring the limits of transfer learning with a unified text-to-text transformer. J. Mach. Learn. Res. **21**(1), 5485–5551 (2020)
4. Radford, A., Narasimhan, K., Salimans, T., Sutskever, I., et al.: Improving language understanding by generative pretraining (2018). https://s3-us-west-2.amazonaws.com/openai-assets/research-covers/language-unsupervised/language_understanding_paper.pdf
5. Radford, A., Wu, J., Child, R., Luan, D., Amodei, D., Sutskever, I., et al.: Language models are unsupervised multitask learners. OpenAI Blog **1**(8), 9 (2019)
6. Brown, T., Mann, B., Ryder, N., Subbiah, M., Kaplan, J.D., Dhariwal, P., Neelakantan, A., Shyam, P., Sastry, G., Askell, A., et al.: Language models are few-shot learners. Adv. Neural Inf. Process. Syst. **33**, 1877–1901 (2020)

Chapter 2
Knowledge Sources

Abstract Knowledge sources are essential components of many NLP tasks, such as question answering (Chen et al., Reading wikipedia to answer open-domain questions. Preprint, 2017), fact verification (Thorne et al., Fever: a large-scale dataset for fact extraction and verification. Preprint, 2018), entity linking (Guo and Barbosa, Semantic Web 9(4):459–479, 2018; Josifoski et al., Zero-shot entity linking with dense entity retrieval. EMNLP, 2020), slot filling (Levy et al., Zero-shot relation extraction via reading comprehension. Preprint, 2017), dialogue (Dinan et al., Wizard of wikipedia: Knowledge-powered conversational agents. Preprint, 2018), etc. One task can also be the knowledge source for another task, such part-of-speech tagging (Schmid, Part-of-speech tagging with neural networks. Preprint, 1994) for dependency parsing (Qi et al., Universal dependency parsing from scratch. Preprint, 2019), etc. However, the availability, quality, and suitability of different types of knowledge sources vary depending on the domain, language, and task requirements. This chapter provides a comprehensive overview of the main types of knowledge sources used in NLP, such as statistical models, knowledge bases, task specific corpus with human annotations, etc.

Keywords Knowledge source · Knowledge base · Large language model · Knowledge from web · Human-annotated knowledge

2.1 Introduction

Natural language processing (NLP) includes various tasks such as machine translation [1], sentiment analysis [17], question answering [2], text summarization [3], etc. To perform these tasks, NLP systems [4] often rely on various types of knowledge sources that can provide relevant information, rules, features, etc. We categorize the knowledge sources into the following aspects.

Pretrained large language models are the most knowledgeable models, such as ELMo [5], BERT [6], T5 [7], GPT models [8], etc. These models are trained on large-scale corpora and can capture various linguistic and semantic patterns. They can be fine-tuned or adapted for specific NLP tasks or domains, used as feature extractors or embeddings, or prompted to get some answers.

Unstructured knowledge source is a collection of texts that are not annotated or organized in a formal way, such as books [9], articles [10], web pages [11], social media posts [12], etc. An unstructured knowledge source can provide diverse and implicit information that can be mined, retrieved, or summarized by NLP methods. For example, an unstructured knowledge source can help to generate natural language texts, to provide background knowledge, or to support open-domain dialogue.

Structured knowledge base [13–15] is a repository of facts, concepts, relations, and entities that are organized in a formal or semi-formal way, such as ontologies, taxonomies, databases, graphs, etc. A structured knowledge base can provide rich and explicit information that can be queried, inferred, or linked to natural language texts. For example, a structured knowledge base can help resolve entity references, enrich text representations, or answer factual questions.

Task-specific corpus is a collection of texts that are annotated with specific labels, such as part-of-speech tags [86], named entities [16, 83, 84], sentiment analysis [17], question answering [2], etc. A human-labeled corpus can provide supervised or semi-supervised data for training or evaluating NLP models, or for extracting features or rules. For example, a human-labeled corpus can help to train a classifier to evaluate a generation model, or to extract syntactic or semantic patterns.

We summarize the different kinds of knowledge resources in Table 2.1, and will introduce more details of different knowledge resources.

2.2   Pretrained Language Models

Pretrained language models (PLM) are computational models that have been trained on large amounts of natural language data in text. The dataset varies from diverse sources and domains, such as Wikipedia, books, news articles, social media posts, etc. The goal of this training is to learn the statistical patterns and regularities of natural language, such as how words, phrases, sentences, and paragraphs are formed, how they relate to each other, and how they convey meaning and context. After pretraining, the models can implicitly encode some knowledge that is relevant for different tasks, such as the syntactic and semantic roles of words and phrases, the common sense and world knowledge that is implied or expressed in the text, the style and tone of the text, etc. This knowledge is distributed and latent in the network's weights and biases, and it is not explicitly labeled or annotated in the training data. There are two kinds of widely adopted pertaining losses.

The auto-regressive language model is a type of generative model that predicts the next word in a sequence based on the previous words. Most of the large-scale language models are pretrained in this direction, such as the series of GPT [8, 11, 18, 19] and LLaMA [20, 20]. Before these large-scale language models with billions of parameters in Transformer [21], LSTM [22] is another popular framework, such as ELMo [5].

Table 2.1 An overview of different knowledge from well-trained models and datasets

Source	Details
Pretrained language models	1. Dataset/Models Large-scale language models: OpenAI GPT, LLaMA, PaLM, Falcon Claude, etc Masked language models: BERT, RoBERTa, BART, T5, etc Domain specific modes: mT5, CodeT5, Legal-BERT, FinBERT, BioBERT, etc.
	2. Usage Model initialization for further fine-tuning or providing sequence embedding. Prompting language models to get knowledge, such as calling OpenAI service. Mask text as query and ask pretrained language model to recover the masked words.
	3. Advantage PLM can learn knowledge from large-scale unlabeled datasets. Query for knowledge is flexible and LLM can have better understanding of the queries. LLM API, such as OpenAI services, is easy to call
	4. Limitation May not be explicit, consistent, accurate, or complete. Computational heavy
Unstructure knowlege	1. Dataset/Models Wikipedia, Books, GitHub, ArXiv, StackExchange, Reddit, Commoncrawl, FreeLaw, Pubmed, EnronEmails
	2. Usage General or domain-specific pertaining with large corpus Corpus of search engines to retrieve external knowledge or text
	3. Advantage No need human annotation, and it contains the largest amount of knowledge. Widely used to train large language models.
	4. Limitation Knowledge or facts from different sources may not be consistent and can be redundant. The data formats are diverse, such as HTML, table, code, etc.
Structure knowledge	1. Dataset/Models Wikidata, DBpedia, YAGO Freebase, ConceptNet, Bio2RDF, GeoNames, WordNet, FrameNet
	2. Usage Model pertaining by inserting related knowledge to text for pretraining. Finding relationships between entities or explanations for specific words.
	3. Advantage. Cleaner knowledge with the human effort involved. Well-defined structure knowledge is easier to understand and retrieve. Well-categorized knowledge, such commonsense knowledge, relation graphl
	4. Limitation Knowledge is always incomplete and human annotation is expensive.

(continued)

Table 2.1 (continued)

Source	Details
Task specific corpus and tool	1. Dataset/Models Instruction tuning, semantic parsing, relation extraction, machine translation, summarization, etc.
	2. Usage Retrieve related labeled data as knowledge Models finetuned on the data can tag external information for other tasks.
	3. Advantage The human-labeled datasets are of high quality. Well-trained tools, like POS, parser, translation, etc, are widely adopted.
	4. Limitation Data annotation is expensive and always incomplete for different domains.

Masked language model is a type of neural network that is trained to predict missing words or tokens based on the surrounding context. For example, BERT [6] randomly masks some tokens in the input text with a special symbol, such as [MASK], and then feeds the masked text to the model. The model then tries to generate the original tokens for the masked positions, using the unmasked tokens as clues. BERT [6], RoBERTa [23], ALBERT [24], treat the task as a sequence labeling problem by using the hidden states on the masked positions to predict the missing word. BART [25], T5 [7] treat the task as a generation problem by using encoder-decoder framework, where the unmasked tokens are encoded first and then the encoding states are used to decode either the whole original sentences or only the masked tokens.

The knowledge learned by pretrained language also depends on the diversity and balance of the training data. Some pretrained language models may have more knowledge about specific domains, such as Codet5 for coding [26], FinBERT [27] for finance, LEGAL-BERT [28] for legal domains, mT5 [29] for multi-lingual domains, etc. There are usually two ways to leverage the knowledge learned from pretraining language models.

Model Initialization with the pretrained models on large and diverse datasets can help the model generalize better to new and related domains or tasks. Some tasks might not have enough high-quality data to train a model from scratch. With the initialization, the model can avoid overfitting or underfitting, and improve its performance and robustness. After the initialization, in order to keep the knowledge learning from pertaining process, there are also different ways to address the issue. One general solution is to fine-tune a smaller number of parameters on the smaller dataset with parameter-efficient tuning. For example, prefix-tuning [30] only optimizes the continuous representation of the prompt, and LoRA [31] adds low-rank weight to the original weights. And we can finetune all the parameters when the data quality is high and have plenty of hardware for model training. There are also works directly using the encoded representation by pretrained model. ELMo [5] is one of the earliest works on building contextual representations.

Prompting Language Model for knowledge is one of the most popular ways now. For the large-scale language models (LLM) aligned with instruction, such as InstructGPT [32], ChatGPT, etc, we can treat them as agents and directly ask questions to collect the knowledge. If the goal is to collect factual knowledge about a specific topic or entity, such as a person, a place, a product, or a concept, we could use ChatGPT to answer questions, provide information, and explain concepts in natural language. For example, we could ask ChatGPT questions like "Who is Elon Musk?", "How does a microwave oven work?", or "What are the benefits of meditation?" and expect to receive relevant and accurate answers from ChatGPT. Alternatively, we could also prompt ChatGPT to initiate a dialogue about a topic or entity of interest by providing keywords and then follow up with more questions to elicit more knowledge. For example, we could start a conversation with ChatGPT by saying, "Tell me about the solar system", and then ask more specific questions or express opinions about the planets, the sun, the asteroids, etc.

While for the pretrained models without instruction tuning, they need more well-designed contextual words to prompt models to provide related knowledge. And the prompting strategies mainly rely on the pertaining loss. For the models trained with masked language modeling (MLM), such as BERT or T5, to generate masked sentences or passages from the corpus, where one or more tokens are replaced with a special mask token, such as [MASK]. We might use the MLM to fill in the masked information to get knowledge, such as "The capital of [MASK] is Berlin" to answer the question, or "Oil prices rise? [MASK] Oil prices fall back" to identify whether the two statements are contradicted [33]. For the models trained with auto-regressive language model, we can directly prompt the model with "The country with Berlin as capital is". The method can also work like a human-annotated knowledge base [34].

Limitations and Advantage Although the PLMs can be used to access factual or commonsense knowledge that is implicitly encoded in the PLM's parameters, this approach may not always be reliable, accurate, or consistent. The PLM may generate incorrect, incomplete, or contradictory responses, depending on the query, the context, and the randomness. Moreover, this approach may not be efficient, as it requires heavy computational cost to generate text for each query. The advantage of PLMs is that they can learn more different knowledge from unlabeled data. PLM can have a better understanding of the question and then generate the answer. With more companies starting to provide LLM API, such as OpenAI, it comes to be easier to use.

2.3   Unstructured Knowledge

An unstructured knowledge base for natural language processing is a collection of texts, documents, or other sources of natural language data that can be used to collect or generate knowledge. The data [35, 36] is widely used for pretraining large language models. The following is about unstructured knowledge base sources.

Wikipedia[1] is a free online encyclopedia. It is one of the most popular sources of information, covering a wide range of topics, from history and science to culture and current events, that can help models gain a basic understanding. Wikipedia can be used as a knowledge base to retrieve relevant information, especially for knowledge-intensive tasks [2, 4, 37, 82, 86].

News article [38, 39] is another common type of natural language content. New articles can provide factual information, such as names, dates, locations, events, statistics, opinions, and quotes, that can be extracted and queried by NLP systems for various applications. Besides, they can also provide feedback and evaluation, such as comments, ratings, and reviews.

Social media platforms such as Twitter, Facebook, Instagram, and Reddit generate large volumes of natural language content, reflecting the opinions, emotions, interests, and behaviors of users. Social media posts can be used for tasks such as sentiment analysis [40], emotion detection [41], topic detection [42], and social network analysis.

Books and literature [9, 43, 44] are rich sources of natural language content, containing stories, characters, themes, and styles. This source can be used for tasks such as text generation, story understanding, character analysis, and style transfer.

Scientific papers and reports [10, 45] present the results, methods, and implications of scientific research. Scientific papers and reports can be used for tasks such as information retrieval, citation analysis, knowledge discovery, and summarization.

Programming languages [46, 47] can provide a rich source of data for training coding models that aim to synthesize, debug, refactor, or document code. For example, code repositories can be used to train coding models for code completion, code synthesis, code summarization, code repair, or code documentation. Some examples of code repositories are GitHub, GitLab, or SourceForge.

We also have different kinds of text from popular domains. Legal texts [28][2] from various sources and domains, such as legislation, case law, contracts, and academic articles. The pretraining data would help with legal question answering, legal summarization, legal reasoning, etc. Biomedical texts [10, 48] cover a broad range of topics and disciplines in the medical domain, and clinical texts[3] use electronic health records from intensive care units (ICUs), containing clinical notes, laboratory results, vital signs, medications, and other information. Financial texts [27] mainly come from financial news articles.

Actually, most of the above sources can come from the Web. The World Wide Web is a vast and diverse source of natural language content, covering various domains, genres, and languages. The easiest way to retrieve knowledge from the web is to call search engines, such as Google, Bing, etc. To have a better understanding of the data, we may not only call search engine APIs, but also have the same analysis on public datasets. Here we would like to list

---

[1] https://dumps.wikimedia.org.

[2] https://www.english-corpora.org/scotus/.

[3] https://physionet.org/content/mimiciii/1.4/.

two resources for data collection, which are for training large language models. The Pile [49][4] provides a downloadable 800G of diverse text. This would be quite useful for research on unstructured knowledge. To be more specific, the dataset includes Pile-CommonCrawl (227.12G, web crawling data), PubMed Central (90.27G, biomedical articles), Books (100.96G, a mix of fiction and nonfiction books), OpenWebText (262.77G, content from Reddit), ArXiv (56.21G, research paper), Github (95.16G, open-source code), FreeLaw (51.15G, legal proceedings), StackExchange (32.20G, user-contributed questions and answers), USPTO Backgrounds (22.90G, US patents), PubMed Abstracts (19.26G, biomedical articles), Gutenberg(PG-19) (10.88G classic Western literature), OpenSubtitles (12.98G, subtitles from movies and television shows), Wikipedia(en) (6.38G, high-quality Wiki text), DMMathematics (7.75G, mathematical problems), UbuntuIRC (5.52G, chatlogs of all Ubuntu-related channels), BookCorpus (26.30G, book), EuroParl (4.59G, multilingual parallel corpus), HackerNews (3.90G, discussion on stories), Youtube Subtitles (3.73G), PhilPapers (2.38G, philosophy publications), NIHExPorter (1.89G, scientific writing), and EnronEmails (0.88G, email). Another popular dataset, RedPajama [50],[5] provides code for crawling LLaMA [20] training dataset. All these datasets can be used for domain-specific knowledge understanding, pertaining, and retrieval.

Limitation and Advantages An unstructured knowledge base may contain noise and inconsistency, which can affect the quality and reliability of the extracted knowledge. It may also require more computational resources and complex algorithms to process and analyze a large amount of natural language data. For example, building an index with dense passage retrieval [37] on all web pages will take many GPU hours. The advantage of unstructured knowledge is that it can capture a rich and diverse range of knowledge that may not be easily formalized or represented in a structured knowledge base. It doesn't need additional human effort to re-label or re-organize the data. It can also adapt to new or emerging knowledge domains, such as posted news and social media information.

2.4 Structured Knowledge

Structured knowledge usually involves human efforts to organize data in specific formats. The most general knowledge bases will link entities, relations, concepts, and facts in a network of nodes and edges. It can be used to answer queries, support reasoning, and enable various applications that require rich and contextual data. Besides, the knowledge structure can also be a dictionary, table, etc.

---

[4] https://pile.eleuther.ai/.

[5] https://github.com/togethercomputer/RedPajama-Data.

Here we first introduce some general-purpose knowledge bases. Note that overlaps exist between different knowledge bases, and Wikidata becomes more popular by merging different sources.

Wikidata[6] is a collaborative, multilingual, open-source knowledge base that integrates data from various sources, such as Wikipedia, Wiktionary, and Wikisource. It provides a common schema and interface for querying and editing. It covers a wide range of topics and entities, and supports logical reasoning and inference. For example, the entity "Douglas Adams", Wikidata shows he is the author of The Hitchhiker's Guide to the Galaxy. Besides, the knowledge base supports various properties and values that describe different aspects of "Douglas Adams", such as date of birth, occupation, nationality, awards, works, influences, and so on. It also has links to other items that are related to Douglas Adams, such as his spouse, his genre, his characters, his adaptations, and his identifiers in other databases. Moreover, each property and value has a source, a qualifier, and a rank to indicate its origin, context, and relevance. The dataset also provides knowledge of common words. For example, for the word "Earth", Wikidata provides its description "third planet from the Sun in the Solar System" and other known names like "Planet Earth", "the world", etc. Wikidata also provides a broad range of statements, such as a picture/video of "earth" and the inception of "earth". There are also different kinds of earlier efforts to construct knowledge based on Wikipedia, such as **DBpedia** [14] and **YAGO** [15].

ConceptNet [13][7] is a knowledge base that collects and organizes common-sense knowledge from various sources, such as Open Mind Common Sense, Wiktionary, and DBpedia. It represents knowledge as a semantic network of nodes and edges, where nodes are concepts and edges are relations, such as IsA, PartOf, or Causes, etc. For example, "a net" is used for "catching fish", "Leaves" is a form of the word "leaf" [13]. It supports natural language understanding and generation, and provides an API and a Python library for querying and manipulating the data.

Freebase[8] is a knowledge base that was created by a community of users and acquired by Google, and contains data about various domains, such as music, movies, sports, and books. It uses a graph-based model and a schema of types and properties to represent entities and facts, and supports reconciliation and disambiguation.

There are also knowledge bases on specific domains. **Bio2RDF**[9] is a biological knowledge base that integrates and interlinks data from various biomedical and life science sources. **GeoNames**[10] is a knowledge base that contains geographical

---

[6] https://www.wikidata.org/wiki/?variant=zh-tw.

[7] https://conceptnet.io/.

[8] https://developers.google.com/freebase.

[9] https://bio2rdf.org/.

[10] https://www.geonames.org/.

information about millions of places, such as names, coordinates, populations, and features. It uses a hierarchical classification of feature codes and a geoname ID to identify and link entities, and supports geocoding and reverse geocoding. It also provides a web service and a dump file for querying and accessing the data.

Next, we introduce some linguistic knowledge bases. **WordNet**[11] is a knowledge base that organizes words into sets of synonyms called synsets, and relates them by semantic and lexical relations, such as hypernyms, hyponyms, antonyms, and meronyms. It covers various parts of speech, such as nouns, verbs, adjectives, and adverbs, and supports lexical analysis and disambiguation. It is available as a database and a software tool for various languages.[12] **FrameNet**[13] is a knowledge base that annotates sentences with semantic frames, which are schematic representations of situations and events, and their participants and roles. It covers various domains and genres, and supports semantic parsing and inference. Here's an example: [(Cook) the boys] . . . GRILL [(Food) their catches] [(Heating_instrument) on an open fire].

Limitation and Advantage   The challenge of structure knowledge base is to design and implement a representation scheme that can capture the relevant and accurate aspects of the domain knowledge. The structure knowledge base can also be huge, and how to elicit, acquire, and validate the knowledge from various sources is challenging. If the whole knowledge base is written by human, it's hard to list a full set. A structured knowledge base can help ensure that the information is accurate and consistent. It is easier and faster to find the related information from knowledge base for specific tasks. The knowledge retrieval process is visualizable and can be used for multi-hop reasoning.

2.5   Task Specific Corpus and Tool

The corpora for solving specific tasks are also valuable resources of expert knowledge. The dataset can be retrieved as additional knowledge [51], and it can also be used to train a tool for labeling, such as the widely used part-of-speech tagging, parsing, etc. There have been many datasets coming out with the growth of artificial intelligence. We will clarify the most popular tasks in the following directions.

Text Generation   has dominated the usage of pre-trained language models, especially after the instruction tuning on LLM. The instruction tuning dataset consists of instructions to solve a problem and the corresponding answers. TULU [52] has a broad analysis of different instruction-tuning datasets. The datasets can be classified

---

[11] https://wordnet.princeton.edu/.

[12] https://www.nltk.org/howto/wordnet.html.

[13] https://framenet.icsi.berkeley.edu/.

into several categories. (1) Transform existing NLP datasets into instruction tuning formats, such as Flan [53] and Super-NaturalInstructions [54]. (2) Human written instructions and responses, such as Dolly [55], OpenAssistant [56], and LIMA [57] (3) Model generated instructions, such as Self-Instruct [58],Alpaca [59] ,Baize [60] (4) User-shared prompts with responses by GPT models, such as ShareGPT [61];

Knowledge Extraction Extracting specific information or entities from a text, such as named entity recognition [16, 62], which identifies the entities or concepts from sentences, and relation extraction [63, 85], which identifies and classifies the semantic relationships between entities. Here's an example: Given the context that "Alice works as a software engineer at Microsoft.", we can extract (1) Alice and software engineer are entities of type person and occupation, respectively. Microsoft is an entity of type organization. (2) Alice and Microsoft are related by the relation "works_at", which indicates an employment relationship. Software engineer and Microfot are related by the relation "occupation_of", which indicates a functional relationship. Semantic parsing [64, 88] will further organize all the entities and relations in structure.

Text Understanding Analyze the meaning and logic of a text on real problems. Natural language inference [65] determines whether a given hypothesis can be logically inferred from a given premise by using the labels entailment, contradiction, or neutral. Sentiment analysis [17, 66, 67] is the process of identifying and extracting the subjective opinions and emotions expressed in a text, speech, or other form of communication. Sentiment analysis can be used for various purposes, such as monitoring customer feedback, reviews, ratings, etc. Topic identification [68] is the process of assigning a label that summarizes the main subject or theme of a text, document, speech, or conversation. It can be used for recommending, personalizing, or filtering information based on the interests, preferences, or needs of users. Question answering [69, 70], especially combined with information retrieval or search engines, can find more specific knowledge.

Text Synthesis Combining or transforming text from different sources. Text simplification [71] is the process of rewriting a text in a way that makes it easier to understand, especially for people with limited literacy skills. Text summarization [72, 73] creates a concise and coherent representation of the main points of a longer text. It can provide an overview or a synopsis of a text for quick browsing or reference, and highlight the key facts or arguments of a text for retrieving external knowledge. Text paraphrasing [74, 75] is the process of rewording or rephrasing a text using different words, expressions, or structures, while preserving its original meaning, tone, and intention. It can increase the text diversity and help the model match knowledge with the same semantics. To further increase the diversity, we can add text style transfer [76, 77], which is the task of transforming a text from one style to another while preserving its meaning.

Multilingual Understanding involves processing, generating, or comparing natural language data in more than one language. Machine translation automatically converts text from one language to another, such as translating a news article from

English to French. Machine translation can facilitate cross-cultural communication, information access, education, tourism, or business. Multilingual natural language understanding is the task in one or more languages, such as identifying the sentiment of a dialogue in English and French.

Domain-Specific Tasks include code completion and math problems. Code completion [78, 79] can be a feature of some programming tools that help programmers write code faster and more accurately by suggesting possible words, symbols, or expressions. Math problems [80, 81] involve solving real problems with basic operations of numbers, such as addition, subtraction, multiplication, and division. Besides these two tasks, any real problem can a domain specific task.

2.6 Conclusion

In this chapter, we have discussed the different types of knowledge sources that can be used for various NLP tasks. We have categorized them into pretrained large language models, unstructured knowledge sources, structured knowledge bases, and task-specific corpora. The latter three sources are more trustworthy, with fewer mistakes or made-up information. They are either human-written or labeled with care. On the other hand, the pretrained large language models can handle the queries for related knowledge better, but they may not produce correct knowledge all the time. Overall, each source has its advantages and limitations, and we can find a suitable source based on the task requirements.

References

1. Zhao, Y., Zhang, J., Zhou, Y., Zong, C.: Knowledge graphs enhanced neural machine translation. In: Proceedings of the Twenty-Ninth International Conference on International Joint Conferences on Artificial Intelligence, pp. 4039–4045 (2021)
2. Chen, D., Fisch, A., Weston, J., Bordes, A.: Reading wikipedia to answer open-domain questions. Preprint (2017). arXiv:1704.00051
3. Gunel, B., Zhu, C., Zeng, M., Huang, X.: Mind the facts: Knowledge-boosted coherent abstractive text summarization. Preprint (2020). arXiv:2006.15435
4. Petroni, F., Piktus, A., Fan, A., Lewis, P., Yazdani, M., De Cao, N., Thorne, J., Jernite, Y., Karpukhin, V., Maillard, J., et al.: Kilt: a benchmark for knowledge intensive language tasks. Preprint (2020). arXiv:2009.02252
5. Iyyer, M., Gardner, M., Clark, C., Lee, K., Zettlemoyer, L., Peters, M.E., Neumann, M.: Deep contextualized word representations. In: North American Chapter of the Association for Computational Linguistics (NAACL) (2018)
6. Devlin, J., Chang, M.-W., Lee, K., Toutanova, K.: Bert: Pre-training of deep bidirectional transformers for language understanding. Preprint (2018). arXiv:1810.04805
7. Raffel, C., Shazeer, N., Roberts, A., Lee, K., Narang, S., Matena, M., Zhou, Y., Li, W., Liu, P.J.: Exploring the limits of transfer learning with a unified text-to-text transformer. J. Mach. Learn. Res. **21**(1), 5485–5551 (2020)

8. OpenAI: Gpt-4 Technical Report, pp. 2303–08774 (2023). arXiv
9. Zhu, Y., Kiros, R., Zemel, R., Salakhutdinov, R., Urtasun, R., Torralba, A., Fidler, S.: Aligning books and movies: towards story-like visual explanations by watching movies and reading books. In: Proceedings of the IEEE international conference on computer vision, pp. 19–27 (2015)
10. Sen, P., Namata, G., Bilgic, M., Getoor, L., Galligher, B., Eliassi-Rad, T.: Collective classification in network data. AI Mag. **29**(3), 93–93 (2008)
11. Radford, A., Wu, J., Child, R., Luan, D., Amodei, D., Sutskever, I., et al.: Language models are unsupervised multitask learners. OpenAI Blog **1**(8), 9 (2019)
12. Hamilton, W., Ying, Z., Leskovec, J.: Inductive representation learning on large graphs. Adv. Neural Inf. Process. Syst. **30** (2017)
13. Speer, R., Chin, J., Havasi, C.: Conceptnet 5.5: an open multilingual graph of general knowledge. In: Proceedings of the AAAI Conference on Artificial Intelligence, vol. 31 (2017)
14. Lehmann, J., Isele, R., Jakob, M., Jentzsch, A., Kontokostas, D., Mendes, P.N., Hellmann, S., Morsey, M., Van Kleef, P., Auer, S., et al.: Dbpedia–a large-scale, multilingual knowledge base extracted from wikipedia. Semantic Web **6**(2), 167–195 (2015)
15. Rebele, T., Suchanek, F., Hoffart, J., Biega, J., Kuzey, E., Weikum, G.: Yago: a multilingual knowledge base from wikipedia, wordnet, and geonames. In: The Semantic Web–ISWC 2016: 15th International Semantic Web Conference, Kobe, Japan, October 17–21, 2016, Proceedings, Part II 15, pp. 177–185. Springer (2016)
16. Nadeau, D., Sekine, S.: A survey of named entity recognition and classification. Lingvisticae Investigationes **30**(1), 3–26 (2007)
17. Zhang, L., Wang, S., Liu, B.: Deep learning for sentiment analysis: a survey. Wiley Interdiscip. Rev. Data Min. Knowl. Discov. **8**(4), e1253 (2018)
18. Radford, A., Narasimhan, K., Salimans, T., Sutskever, I., et al.: Improving language understanding by generative pre-training. Preprint (2018)
19. Brown, T., Mann, B., Ryder, N., Subbiah, M., Kaplan, J.D., Dhariwal, P., Neelakantan, A., Shyam, P., Sastry, G., Askell, A., et al.: Language models are few-shot learners. Adv. Neural Inf. Process. Syst. **33**, 1877–1901 (2020)
20. Touvron, H., Martin, L., Stone, K., Albert, P., Almahairi, A., Babaei, Y., Bashlykov, N., Batra, S., Bhargava, P., Bhosale, S., et al.: Llama 2: open foundation and fine-tuned chat models. Preprint (2023). arXiv:2307.09288
21. Vaswani, A., Shazeer, N., Parmar, N., Uszkoreit, J., Jones, L. Gomez, A.N., Kaiser, Ł., Polosukhin, I.: Attention is all you need. Adv. Neural Inf. Process. Syst.. **30** (2017)
22. Hochreiter, S., Schmidhuber, J.: Long short-term memory. Neural Comput. **9**(8), 1735–1780 (1997)
23. Liu, Y., Ott, M., Goyal, N., Du, J., Joshi, M., Chen, D., Levy, O., Lewis, M., Zettlemoyer, L., Stoyanov, V.: Roberta: A robustly optimized bert pretraining approach. Preprint (2019). arXiv:1907.11692
24. Lan, Z., Chen, M., Goodman, S., Gimpel, K., Sharma, P., Soricut, R.: Albert: a lite bert for self-supervised learning of language representations. Preprint (2019). arXiv:1909.11942
25. Lewis, M., Liu, Y., Goyal, N., Ghazvininejad, M., Mohamed, A., Levy, O., Stoyanov, V., Zettlemoyer, L.: Bart: Denoising sequence-to-sequence pre-training for natural language generation, translation, and comprehension. Preprint (2019). arXiv:1910.13461
26. Wang, Y., Wang, W., Joty, S., Hoi, S.C.H.: Codet5: Identifier-aware unified pre-trained encoder-decoder models for code understanding and generation. Preprint (2021). arXiv:2109.00859
27. Araci, D.: Finbert: financial sentiment analysis with pre-trained language models. Preprint (2019). arXiv:1908.10063
28. Chalkidis, I., Fergadiotis, M., Malakasiotis, P., Aletras, N., Androutsopoulos, I.: Legal-bert: the muppets straight out of law school. Preprint (2020). arXiv:2010.02559
29. Xue, L., Constant, N., Roberts, A., Kale, M., Al-Rfou, R., Siddhant, A., Barua, A., Raffel, C.: mt5: a massively multilingual pre-trained text-to-text transformer. Preprint (2020). arXiv:2010.11934

30. Lisa Li, X., Liang, P.: Prefix-tuning: optimizing continuous prompts for generation. Preprint (2021). arXiv:2101.00190
31. Hu, E.J., Shen, Y., Wallis, P., Allen-Zhu, Z., Li, Y., Wang, S., Wang, L., Chen, W.: Lora: low-rank adaptation of large language models. Preprint (2021). arXiv:2106.09685
32. Ouyang, L., Wu, J., Jiang, X., Almeida, D., Wainwright, C., Mishkin, P., Zhang, C., Agarwal, S., Slama, K., Ray, A., et al.: Training language models to follow instructions with human feedback. Adv. Neural Inf. Process. Syst. (2022)
33. Schick, T., Schütze, H.: It's not just size that matters: Small language models are also few-shot learners. Preprint (2020). arXiv:2009.07118
34. Petroni, F., Rocktäschel, T., Lewis, P., Bakhtin, A., Wu, Y., Miller, A.H., Riedel, S.: Language models as knowledge bases? Preprint (2019). arXiv:1909.01066
35. Soldaini, L., Lo, K., Kinney, R., Naik, A., Ravichander, A., Bhagia, A., Groeneveld, D., Schwenk, D., Magnusson, I., Chandu, K.: Dolma: an open corpus of three trillion tokens for language model pretraining research. Preprint (2024). arXiv.:2402.00159
36. Gao, L., Biderman, S., Black, S., Golding, L., Hoppe, T., Foster, C., Phang, J., He, H., Thite, A., Nabeshima, N., Presser, S., Leahy, C.: The Pile: An 800gb dataset of diverse text for language modeling. Preprint (2020). arXiv:2101.00027
37. Karpukhin, V., Oğuz, B., Min, S., Lewis, P., Wu, L., Edunov, S., Chen, D., Yih, W.-T.: Dense passage retrieval for open-domain question answering. Preprint (2020). arXiv:2004.04906
38. Zhang, J., Zhao, Y., Saleh, M., Liu, P.: Pegasus: Pre-training with extracted gap-sentences for abstractive summarization. In: International Conference on Machine Learning (ICML) (2020)
39. Mackenzie, J., Benham, R., Petri, M., Trippas, J.R., Culpepper, J.S., Moffat, A.: Cc-news-en: a large english news corpus. In Proceedings of the 29th ACM International Conference on Information & Knowledge Management, pp. 3077–3084 (2020)
40. Kharde, V., Sonawane, P., et al.: Sentiment analysis of twitter data: a survey of techniques. Preprint (2016). arXiv:1601.06971
41. Colnerič, N., Demšar, J.: Emotion recognition on twitter: comparative study and training a unison model. IEEE Trans. Affect. Comput. **11**(3), 433–446 (2018)
42. Weng, J., Lim, E.-P., Jiang, J., He, Q.: Twitterrank: finding topic-sensitive influential twitterers. In: Proceedings of the third ACM International Conference on Web Search and Data Mining, pp. 261–270 (2010)
43. Hill, F., Bordes, A., Chopra, S., Weston, J.: The goldilocks principle: reading children's books with explicit memory representations. Preprint (2015). arXiv:1511.02301
44. Wang, Y., Le, H., Gotmare, A.D., Bui, N.D.Q., Li, J., Hoi, S.C.H.: Codet5+: open code large language models for code understanding and generation. Preprint (2023). arXiv:2305.07922
45. Clement, C.B., Bierbaum, M., O'Keeffe, K.P., Alemi, A.A.: On the use of arxiv as a dataset. Preprint (2019). arXiv:1905.00075
46. Puri, R., Kung, D.S., Janssen, G., Zhang, W., Domeniconi, G., Zolotov, V., Dolby, J., Chen, J., Choudhury, M., Decker, L., et al.: Codenet: a large-scale ai for code dataset for learning a diversity of coding tasks. Preprint (2021). arXiv:2105.12655
47. Zhang, J., Panthaplackel, S., Nie, P., Li, J.J., Gligoric, M.: Coditt5: Pretraining for source code and natural language editing. In: Proceedings of the 37th IEEE/ACM International Conference on Automated Software Engineering, pp. 1–12 (2022)
48. Lee, J., Yoon, W., Kim, S., Kim, D., Kim, S., So, C.H., Kang, J.: Biobert: a pre-trained biomedical language representation model for biomedical text mining. Bioinformatics **36**(4), 1234–1240 (2020)
49. Gao, L., Biderman, S., Black, S., Golding, L., Hoppe, T., Foster, C., Phang, J., He, H., Thite, A., Nabeshima, N., et al.: The pile: an 800gb dataset of diverse text for language modeling. Preprint (2020). arXiv:2101.00027
50. Together Computer: Redpajama: an open source recipe to reproduce llama training dataset. Preprint (2023)
51. Wang, S., Xu, Y., Fang, Y., Liu, Y., Sun, S., Xu, R., Zhu, C., Zeng, M.: Training data is more valuable than you think: a simple and effective method by retrieving from training data. Preprint (2022). arXiv:2203.08773

52. Wang, Y., Ivison, H., Dasigi, P., Hessel, J., Khot, T., Chandu, K.R., Wadden, D., MacMillan, K., Smith, N.A., Beltagy, I., et al.: How far can camels go? exploring the state of instruction tuning on open resources. Preprint (2023). arXiv:2306.04751

53. Longpre, S., Hou, L., Vu, T., Webson, A., Chung, H.W., Tay, Y., Zhou, D., Le, Q.V., Zoph, B., Wei, J., et al.: The flan collection: designing data and methods for effective instruction tuning. Preprint (2023). arXiv:2301.13688

54. Wang, Y., Mishra, S., Alipoormolabashi, P., Kordi, Y., Mirzaei, A., Arunkumar, A., Ashok, A., Dhanasekaran, A.S., Naik, A., Stap, D., et al.: Super-naturalinstructions: generalization via declarative instructions on 1600+ nlp tasks. Preprint (2022). arXiv:2204.07705

55. Conover, M., Hayes, M., Mathur, A., Meng, X., Xie, J., Wan, J., Shah, S., Ghodsi, A., Wendell, P., Zaharia, M., et al.: Free dolly: introducing the world's first truly open instruction-tuned llm. Preprint (2023)

56. Köpf, A., Kilcher, Y., von Rütte, D., Anagnostidis, S., Tam, Z.-R., Stevens, K., Barhoum, A., Duc, N.M., Stanley, O., Nagyfi, R., et al.: Openassistant conversations–democratizing large language model alignment. Preprint (2023). arXiv:2304.07327

57. Zhou, C., Liu, P., Xu, P., Iyer, S., Sun, J., Mao, Y., Ma, X., Efrat, A., Yu, P., Yu, L., et al.: Lima: less is more for alignment. Preprint (2023). arXiv:2305.11206

58. Wang, Y., Kordi, Y., Mishra, S., Liu, A., Smith, N.A., Khashabi, D., Hajishirzi, H.: Self-instruct: aligning language model with self generated instructions. Preprint (2022). arXiv:2212.10560

59. Taori, R., Gulrajani, I., Zhang, T., Dubois, Y., Li, X., Guestrin, C., Liang, P., Hashimoto, T.B.: Stanford alpaca: an instruction-following llama model. Github, San Francisco (2023)

60. Xu, C., Guo, D., Duan, N., McAuley, J.: Baize: An open-source chat model with parameter-efficient tuning on self-chat data. Preprint (2023). arXiv:2304.01196

61. Chiang, W.-L., Li, Z., Lin, Z., Sheng, Y., Wu, A., Zhang, H., Zheng, L., Zhuang, S., Zhuang, Y., Gonzalez, J.E., et al.: Vicuna: an open-source chatbot impressing gpt-4 with 90%* chatgpt quality. See https://vicuna.lmsys.org(Accessed 14 April 2023) (2023)

62. Li, J., Sun, A., Han, J., Li, C.: A survey on deep learning for named entity recognition. IEEE Trans. Knowl. Data Eng. **34**(1), 50–70 (2020)

63. Zhou, G., Su, J., Zhang, J., Zhang, M.: Exploring various knowledge in relation extraction. In: Proceedings of the 43rd Annual Meeting of the Association for Computational Linguistics (acl'05), pp. 427–434 (2005)

64. Berant, J., Chou, A., Frostig, R., Liang, P.: Semantic parsing on freebase from question-answer pairs. In: Proceedings of the 2013 Conference on Empirical Methods in Natural Language Processing, pp. 1533–1544 (2013)

65. Bowman, S.R., Angeli, G., Potts, C., Manning, C.D. A large annotated corpus for learning natural language inference. Preprint (2015). arXiv:1508.05326

66. Socher, R., Perelygin, A., Wu, J., Chuang, J., Manning, C.D., Ng, A.Y., Potts, C: Recursive deep models for semantic compositionality over a sentiment treebank. In: Proceedings of the 2013 Conference on Empirical Methods in Natural Language Processing, pp. 1631–1642 (2013)

67. Maas, A., Daly, R.E., Pham, P.T., Huang, D., Ng, A.Y., Potts, C.: Learning word vectors for sentiment analysis. In: Proceedings of the 49th Annual Meeting of the Association for Computational Linguistics: Human Language Technologies, pp. 142–150 (2011)

68. Zhang, X., Zhao, J., LeCun, Y.: Character-level convolutional networks for text classification. Adv. Neural Inf. Process. Syst. **28** (2015)

69. Rajpurkar, P., Zhang, J., Lopyrev, K., Liang, P.: Squad: 100,000+ questions for machine comprehension of text. Preprint (2016). arXiv:1606.05250

70. Kwiatkowski, T., Palomaki, J., Redfield, O., Collins, M., Parikh, A., Alberti, C., Epstein, D., Polosukhin, I., Devlin, J., Lee, K., et al.: Natural questions: a benchmark for question answering research. Trans. Assoc. Comput. Linguist. **7**, 453–466 (2019)

71. Martin, L., Fan, A., de La Clergerie, E.V., Bordes, A., Sagot, B.: Multilingual unsupervised sentence simplification (2021). arXiv:2005.00352

72. Rush, A.M., Chopra, S., Weston, J.: A neural attention model for abstractive sentence summarization. Preprint (2015). arXiv:1509.00685
73. Nallapati, R., Zhou, B., Gulcehre, C., Xiang, B., et al.: Abstractive text summarization using sequence-to-sequence rnns and beyond. Preprint (2016). arXiv:1602.06023
74. Fader, A., Zettlemoyer, L., Etzioni, O.: Paraphrase-driven learning for open question answering. In: Proceedings of the 51st Annual Meeting of the Association for Computational Linguistics (Volume 1: Long Papers), pp. 1608–1618 (2013)
75. Lin, T.-Y., Maire, M., Belongie, S., Hays, J., Perona, P., Ramanan, D., Dollár, P., Zitnick, C.L.: Microsoft coco: common objects in context. In: Computer Vision–ECCV 2014: 13th European Conference, Zurich, Switzerland, September 6-12, 2014, Proceedings, Part V 13, pp. 740–755. Springer (2014)
76. Fu, Z., Tan, X., Peng, N., Zhao, D., Yan, R.: Style transfer in text: exploration and evaluation. In Proceedings of the AAAI Conference on Artificial Intelligence, vol. 32 (2018)
77. Krishna, K., Wieting, J., Iyyer, M.: Reformulating unsupervised style transfer as paraphrase generation. Preprint (2020). arXiv:2010.05700
78. Raychev, V., Vechev, M., Yahav, E.: Code completion with statistical language models. In: Proceedings of the 35th ACM SIGPLAN Conference on Programming Language Design and Implementation, pp. 419–428 (2014)
79. Chen, M., Tworek, J., Jun, H., Yuan, Q., de Oliveira Pinto, H.P., Kaplan, J., Edwards, H., Burda, Y., Joseph, N., Brockman, G., Ray, A., Puri, R., Krueger, G., Petrov, M., Khlaaf, H., Sastry, G., Mishkin, P., Chan, B., Gray, S., Ryder, N., Pavlov, M., Power, A., Kaiser, L., Bavarian, M., Winter, C., Tillet, P., Such, F.P., Cummings, D., Plappert, M., Chantzis, F., Barnes, E., Herbert-Voss, A., Guss, W.H., Nichol, A., Paino, A., Tezak, N., Tang, J., Babuschkin, I., Balaji, S., Jain, S., Saunders, W., Hesse, C., Carr, A.N., Leike, J., Achiam, J., Misra, V., Morikawa, E., Radford, A., Knight, M., Brundage, M., Murati, M., Mayer, K., Welinder, P., McGrew, B., Amodei, D., McCandlish, S., Sutskever, I., Zaremba, W.: Evaluating large language models trained on code. Preprint (2021). arXiv: 2107.03374
80. Hendrycks, D., Burns, C., Kadavath, S., Arora, A., Basart, S., Tang, E., Song, D., Steinhardt, J.: Measuring mathematical problem solving with the math dataset. In: NeurIPS (2021)
81. Cobbe, K., Kosaraju, V., Bavarian, M., Chen, M., Jun, H., Kaiser, L., Plappert, M., Tworek, J., Hilton, J., Nakano, R., Hesse, C., Schulman, J.: Training verifiers to solve math word problems. Preprint (2021). arXiv:2110.14168
82. Thorne, J., Vlachos, A., Christodoulopoulos, C., Mittal, A.: Fever: a large-scale dataset for fact extraction and verification. Preprint (2018). arXiv:1803.05355
83. Guo, Z., Barbosa, D.: Robust named entity disambiguation with random walks. Semantic Web 9(4), 459–479 (2018)
84. Josifoski, M., Riedel, S., Zettlemoyer, L., Wu, L., Petroni, F.: Zero-shot entity linking with dense entity retrieval. In: EMNLP (2020)
85. Levy, O., Seo, M., Choi, E., Zettlemoyer, L.: Zero-shot relation extraction via reading comprehension. Preprint (2017). arXiv:1706.04115
86. Dinan, E., Roller, S., Shuster, K., Fan, A., Auli, M., Weston, J.: Wizard of wikipedia: Knowledge-powered conversational agents. Preprint (2018). arXiv:1811.01241
87. Schmid, H.: Part-of-speech tagging with neural networks. Preprint (1994). arXiv:cmp-lg/9410018
88. Qi, P., Dozat, T., Zhang, Y., Manning, C.D.: Universal dependency parsing from scratch. Preprint (2019). arXiv:1901.10457

Chapter 3
Knowledge-augmented Methods for Natural Language Understanding

Abstract This chapter delves into the emerging domain of knowledge-augmented natural language understanding (NLU), an essential aspect of natural language processing. The integration of external knowledge sources with pretrained language models is key to tackling a wide range of NLU tasks, including question answering, fact verification, and knowledge graph-based tasks. This chapter systematically explores the three steps of knowledge integration—representation, grounding, and integration, with analysis of both structured and unstructured knowledge sources. Through an examination of state-of-the-art models and techniques, the chapter offers a close look at recent advances in knowledge-augmented NLU domain. This comprehensive overview aims to equip readers with a deeper understanding of the current landscape and future potential of knowledge-augmented NLU.

Keywords Natural language understanding · Knowledge-augmented language models · Open domain question answering · Entity linking · Link prediction · Fact verification

3.1 Introduction

In natural language understanding (NLU), the task is to make predictions about the property of words, phrases, sentences or paragraphs based on the input text, e.g., sentiment analysis, named entity recognition and language inference. In domains and tasks such as fact verification and question answering, additional context is necessary to correctly understand the question and make predictions. These tasks represent a unique challenge for the traditional pretrain-and-finetune framework that uses a pretrained language model to solve NLU tasks. A model will have to rely on its own world knowledge to solve the given task, but it is almost impossible to store all the knowledge into a single model.

To alleviate the problem, models and methods have been proposed to integrate external knowledge sources into a language model. These models can use either structured or unstructured knowledge(c.f., Chap. 2). A model typically operates in 3 steps: (1) Converting the knowledge into a manageable representation (knowledge

M. Jiang et al., *Knowledge-augmented Methods for Natural Language Processing*, SpringerBriefs in Computer Science, https://doi.org/10.1007/978-981-97-0747-8_3

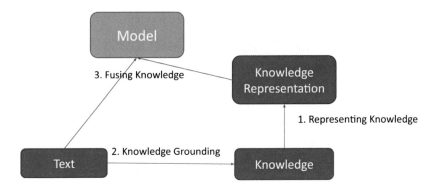

Fig. 3.1 Graphical illustration of the three steps in knowledge-augmented NLU

representation); (2) (Map the given input to relevant knowledge in the knowledge source (knowledge grounding); (3) Integrate the given knowledge into a language model (knowledge integration). Figure 3.1 illustrates the three steps in knowledge-augmented NLU.

In this chapter, we introduce models and methods for knowledge-augmented NLU. We first introduce NLU tasks that require external knowledge in Sect. 3.2. We will introduce how to use knowledge to augment NLU models along the 3 dimensions above for both structured and unstructured knowledge (Sect. 3.3). We introduce representative papers in this area in Sect. 3.5, and give a conclusion and future directions in Sect. 3.6.

3.2 Tasks and Benchmarks

It is common for humans to resort to external knowledge sources to perform language understanding tasks. Prior research has established numerous tasks and benchmarks to test a model's ability to use external knowledge to understand language. Here we describe major tasks in this area.

3.2.1 Question Answering

In Question Answering (QA), the machine learning system is required to take a given question from the user and answer the question with its knowledge. QA naturally requires external knowledge as it often requires retrieving and reasoning over information that is not present in the question. Two major types of QA require external knowledge: Open-domain QA, and Commonsense QA.

Open-Domain Question Answering (ODQA) requires the machine to answer a question that requires external information to answer. For example, a question from Natural Questions [1] is

– Q: *"What was the first capital city of Australia?"*
– A: *"Melbourne"*.

Natural Questions [1] is one of the most popular datasets for ODQA. It has 307,373 questions from real Google search queries. Every question is annotated with a long answer (about the length of a paragraph) and a short answer (one or more entities), along with the top 5 Wikipedia pages from the search result. In the Open-domain setting, the goal is usually to generate a short answer without relying on Wikipedia pages. Besides Natural Questions, other popular ODQA datasets include TriviaQA [2], ELI5 [3], and HotpotQA [4].

Besides pure text-based QA, other datasets aim to extract question-answer from knowledge graphs. These datasets in the **Knowledge Base Question Answering (KBQA)** domain are also good testbeds for ODQA if the machine does not rely on the same knowledge base to construct the datasets. For example, WebQuestions [5] extracts question-answer pairs based on Freebase. Subsequent papers use Wikipedia or other knowledge graphs (e.g., Wikidata) to solve WebQuestions in the open domain.

Commonsense Question Answering is another domain of QA that focuses on understanding the common sense knowledge in the models. Different from ODQA which typically relies on *explicit* external knowledge like Wikipedia, Commonsense QA focuses on the *implicit* common sense and background knowledge for a given question. For example, a question from the CommonsenseQA dataset [6] is

Q: *"What do all humans want to experience in their own home?"*
(A) feel comfortable, (B) work hard, (C) fall in love, (D) lay eggs, (E) live forever".

Commonsense QA has seen the development of various datasets, each designed to probe different facets of commonsense reasoning. One prominent example is the CommonsenseQA dataset [6], which focuses on General commonsense knowledge based on ConceptNet [7]. Other noteworthy datasets in this domain include SocialIQA [8], which focuses on social interactions and relationships, and PhysicalIQA [9], aimed at evaluating a system's understanding of physical world phenomena.

While some question answering tasks like machine comprehension [10] are designed to not rely on external information, integrating external knowledge can still help boost the performance on such tasks [11].

3.2.2   Knowledge-Graph Related Tasks

Knowledge graphs, such as Freebase and Wikidata, have a wide application in search engines, chatbots, and recommendation systems. Knowledge in knowledge graphs is typically stored in the formats of (subject, relation, object) triplets. For example, (United States, *IsA*, Country). Knowledge graphs provide structured representations of facts, entities and relationships that facilitate reasoning over knowledge. Tasks for knowledge graphs closely follow their applications. In addition to Knowledge Base QA that we already covered in Sect. 3.2.1, we introduce two other tasks for knowledge graphs in this section.

Entity Linking (EL) refers to associating entity mentions in a given text to the corresponding entities in a knowledge graph (e.g., Wikidata entries or Wikipedia pages). EL is crucial for grounding text mentions in knowledge graphs and is the first step in using knowledge graphs for various applications like search engines and chatbots. A prominent dataset in EL is WNED-Wiki [12], a dataset automatically created by sampling documents from the 2013 Wikipedia dump and balancing the difficulty of linking each mention using a baseline as a proxy.

Slot Filling/Link Prediction involves predicting links (triplets) in the knowledge graph. In its most common form, the task is to predict the object given subject and the relation. Knowledge graphs are far from complete, so using ML to infer the missing facts can dramatically reduce manual curation requirements. The ability to understand missing links also indicates a model's ability to understand and reason over the knowledge graph. A recent big dataset in this area is the WikiKG90Mv2 [13] dataset extracted from the entire Wikidata knowledgebase, with 90M entities and 600M triplets in total in its newest version. Another relevant dataset is the LAMA probe [14], which tests a language model by autocompletion to test its internal factual and commonsense knowledge. It uses knowledge from 4 different sources including wikidata, Wikipedia and commonsense knowledge in ConceptNet.

3.2.3   Knowledge-Related Tasks in Other Applications

Besides question answering and knowledge graph-related tasks, external knowledge is indispensable for a variety of language understanding tasks derived from downstream applications.

Fact Verification refers to assessing the validity of a given claim based on public knowledge sources. Given the quick spread of information and generative models' ability to generate human-like text, the demand for a good fact verification system is rapidly increasing. FEVER [15] is a commonly used dataset for fact verification. FEVER contains 185,445 claims generated by altering claims extracted from Wikipedia, and then verified by annotators. It requires retrieving sentence-level evidence to support if a claim is supported or refuted.

Dialogue Response Generation is a classic task in dialogue systems to predict a good reply for a given dialogue. Using external knowledge for open-domain dialogues is instrumental in generating interesting and inspiring responses. Wizard of Wikipedia [16] is a large-scale dialogue dataset focusing on external knowledge. It has 20k conversations grounded with knowledge retrieved from Wikipedia. The goal is to generate the next response using knowledge in Wikipedia.

Commonsense Knowledge Tasks People have also proposed various tasks to understand a model's ability to understand common sense knowledge. The Winograd Schema Challenge [17] uses coreference resolution as a proxy for common sense knowledge probes. Understanding the coreference ambiguity in the text usually requires a substantial amount of common sense. Hellaswag [18] proposes to use next sentence prediction to probe a model's capability for common sense. The machine is required to select a proper continuation from 4 candidates for a given context.

Benchmarks Related to External Knowledge KILT [19] is a collection of benchmarks of many of the tasks above, for developing generic models for knowledge-rich tasks. It consists of 5 types of tasks (Open-domain QA, Slot filling, dialogue, fact-checking and entity linking) from 11 datasets. Another important generic benchmark is BEIR [20], which focuses on the information retrieval side of knowledge. It consists of 18 public datasets, and the goal is to retrieve the corresponding knowledge for the given input.

3.3 Models and Methods for Structured Knowledge

Now we introduce models and methods for structured knowledge like knowledge graphs, and semi-structured knowledge like dictionaries.

3.3.1 Knowledge Representation

A straightforward way to represent the knowledge in knowledge graphs is to use existing knowledge embeddings for link prediction tasks. For example, ERNIE [21] uses TransE representations [22] and directly incorporates the embeddings into a language model. Knowledge graph embeddings have a long history and have seen extensive research through the years. A survey of knowledge graph embeddings is out of the scope of this book, and we refer authors to the survey of [23] for an extensive review.

On the other hand, knowledge graph embeddings are typically developed for the link prediction task in mind, which is not necessarily the same requirement for knowledge-rich NLU tasks. Training the knowledge embeddings together with the language models can lead to embeddings that are more helpful for understanding the context and thus more suitable for NLU. Entity-as-Experts (EAE) [24] jointly trains the embeddings with a language model using the traditional masked language modeling loss, along with entity linking and mention detection loss. The resulting model has a substantial improvement in the LAMA probe and open-domain QA tasks. FILM [25] additionally utilizes knowledge graph links as a "facts memory" to further improve performance. Oreo-LM [26] formalizes reasoning over knowledge graphs as a random walk over nodes and extends the idea to generative models.

Another way to represent knowledge is to utilize the definition or description that comes together with the knowledge graph. For example, KEPLER[27] uses RoBERTa [28] to process the entity descriptions and use the embedding of the $<s>$ token as the entity embedding. One problem of this method is the coverage—only a small fraction of nodes in Wikidata have descriptions, and many other knowledge bases do not have a dedicated description field. Dictionaries or other definition-focused databases are good alternatives for this purpose. Dict-BERT [29] uses dictionary definitions to help understand rare words in NLU tasks.

3.3.2 Knowledge Grounding

Since KGs have clearly defined structures, grounding the input text into KG just requires entity linking to find the corresponding entity in the KG. Many papers simply use heuristic and string matching to link the entity [29–31]. Others use existing entity linking tools for better quality linking. For example, ERNIE [21] uses TAGME [32] to identify entities in any given input. Another way is to utilize metadata from text to identify entities. For example, EaE [24] uses hyperlinks in Wikipedia to identify entity mentions in Wikipedia pages.

Once the text is connected to entities in the knowledge graph (or other sources like dictionaries), simple heuristics can retrieve relevant facts using the edges in KG. Methods like ERNIE [21] and KEAR [33] use links between entity mentions as additional information. KG-based methods like Oreo-LM [26] and JAKET [30] retrieve 2 or 3 hop information in the knowledge graph neighborhood for better knowledge coverage.

3.3.3 Knowledge Fusion

Methods for fusing structured knowledge into LMs mostly fall into two categories: (1) fine-tuning methods, which finetune a pretrained LM to use external knowledge; (2) pretraining methods, which jointly train a language model with external knowledge.

Fine-Tuning Methods has the advantage of simplicity: It usually just requires several rounds of fine-tuning on downstream tasks. The simplest way is to linearize the external knowledge as text and append it to the context. This method is simple and effective, and already improves performance on a lot of tasks [31, 34]. One can also train on knowledge-graph tasks directly; for example, GENRE [35] finetunes pretrained language models on the entity linking task by letting it generate Wikipedia titles directly. The resulting model has improved performance for knowledge retrieval tasks and the KILT benchmark.

Pretraining Methods involve more heavy-weight training, but usually provides the language model a better ability to use the external knowledge. For example, JAKET [30] proposes to use the knowledge module to produce embeddings for entities in the text while using the language module to generate context-aware initial embeddings for entities and relations in the knowledge graph. Dict-BERT [36] uses dictionary definitions to help train the language modeling task.

3.4 Models and Methods for Unstructured Knowledge

Unstructured knowledge sources like Wikipedia, news articles, and academic papers don't have an explicit way of grounding and reasoning over knowledge: the knowledge is usually scattered around paragraphs. Therefore, most methods for using unstructured external knowledge resort to some kind of retrieval methods, with discrete or continuous representations. We introduce such methods in this section.

While this section focuses on unstructured knowledge, we note that a model is not constrained to use a single source of knowledge. Using both structured and unstructured knowledge can generally further boost the performance of a language model [37].

3.4.1 Knowledge Representation

Unstructured knowledge representation centers around the representation of text. One way is to rely on traditional discrete searches like Elastic Search or BM25 [38], using the input as a query. For example, DrQA [39] uses TF-IDF vectors and bigrams to retrieve relevant passages for a given question in ODQA. REINA [40] uses BM25 to retrieve relevant training examples for a given test question from a collection of training data.

The unstructured nature of text can make the traditional matching-based discrete method less suitable for knowledge-rich NLU tasks. To properly retrieve the correct information, it is critical that the retriever can "understand" both the query and the knowledge base. Therefore, continuous representations, or document embeddings, can lead to better retrieval performance. DPR [41] forms the foundation of most document embedding methods. It uses a text encoder to map both queries and documents into continuous vectors. As an extension, ColBERT [42] encodes every word in the document as a contextual vector for retrieval. This increases the size of the vector database but can lead to better retrieval performance.

3.4.2   Knowledge Grounding

Once passages are mapped to vectors, the knowledge grounding process needs to retrieve the corresponding knowledge for the query. Methods like DrQA [39] maps questions and passage using the same method (TF-IDF, bigrams), and uses feature hashing [43] to improve the retrieval efficiency.

For continuous vectors, DPR [41] uses a contrastive loss to train the text encoder as the retriever. The query and document embeddings are matched using cosine similarity. At inference time, they use FAISS [44] to efficiently retrieve relevant passages for a given question. RETRO [45] instead simply use a frozen BERT encoder to encode text chunks for training language models.

3.4.3   Knowledge Fusion

The simplest way to incorporate unstructured knowledge is to append them as input to the language model [31, 34, 41, 46]. The self-attention mechanism can enable the language model to reason between the external knowledge and the given input.

A drawback is the input length limit: the input length can increase quickly as more retrieved passages are added. As traditional self-attention has a quadratic dependence on the input length, this increases the running time and memory requirement significantly. To alleviate this issue, [47] proposes the Fusion-in-Decoder model using the encoder-decoder architecture. Instead of joining all passages together as input, FiD encodes every passage separately with the query in the encoder and then fuses them together in the decoder. This greatly reduces the burden on the encoder.

3.5   Representative Models

In this section, we introduce some representative models that utilizes knowledge in NLU. We cover methods focusing on structured knowledge (Sects. 3.5.1, 3.5.2, and 3.5.3), unstructured knowledge (Sects. 3.5.4, 3.5.5, 3.5.6, and 3.5.8), and methods fusing both kinds of knowledge (Sect. 3.5.7).

3.5.1 ERNIE

ERNIE [21] proposes to improve traditional language models by incorporating knowledge graph embeddings. Given a input token sequence $\{w_1, \ldots, w_n\}$, it first uses TAGME [32] to identify the entities $\{e_1, \ldots, e_m\}$ in the input. ERNIE consists of a text encoder (T-Encoder) for encoding lexical and syntactical information, and a knowledge encoder (K-Encoder) to integrate external knowledge. Given any input, $\{w_1, \ldots, w_n\}$ first passes through T-Encoder to get contextual representations $\{w_1^o, \ldots, w_n^o\}$. Entities are converted into embeddings $\{e_1, \ldots, e_m\}$ using pretrained TransE embeddings.

In i-th layer of K-Encoder, the current embeddings $\{w_1^{(i-1)}, \ldots, w_n^{(i-1)}\}$ and $\{e_1^{(i-1)}, \ldots, e_m^{(i-1)}\}$ first goes through a multi-head attention layer:

$$\{\tilde{w}_1^{(i)}, \ldots, \tilde{w}_n^{(i)}\} = \mathtt{MH-ATT}\{w_1^{(i-1)}, \ldots, w_n^{(i-1)}\}, \tag{3.1}$$

$$\{\tilde{e}_1^{(i)}, \ldots, \tilde{e}_m^{(i)}\} = \mathtt{MH-ATT}\{e_1^{(i-1)}, \ldots, e_m^{(i-1)}\}. \tag{3.2}$$

Then the entity embedding is aggregated with the corresponding word. Suppose word w_j is mapped with entity e_k, the aggregation is

$$\mathbf{h}_j = \sigma\left(\widetilde{\mathbf{W}}_t^{(i)}\tilde{\mathbf{w}}_j^{(i)} + \widetilde{\mathbf{W}}_e^{(i)}\tilde{\mathbf{e}}_k^{(i)} + \widetilde{b}^{(i)}\right), \tag{3.3}$$

$$\mathbf{w}_j^{(i)} = \sigma\left(\mathbf{W}_t^{(i)}\mathbf{h}_j + b_t^{(i)}\right), \tag{3.4}$$

$$\mathbf{e}_k^{(i)} = \sigma\left(\mathbf{W}_e^{(i)}\mathbf{h}_j + b_e^{(i)}\right). \tag{3.5}$$

If a word does not have a corresponding entity, its embedding is

$$\mathbf{h}_j = \sigma\left(\widetilde{\mathbf{W}}_t^{(i)}\tilde{\mathbf{w}}_j^{(i)} + \widetilde{b}^{(i)}\right), \tag{3.6}$$

$$\mathbf{w}_j^{(i)} = \sigma\left(\mathbf{W}_t^{(i)}\mathbf{h}_j + b_t^{(i)}\right). \tag{3.7}$$

In order to train ERNIE, [21] propose a new training task that randomly masks entities in text and asks the model to predict them. Given the word representations $\{\mathbf{w}_1, \ldots, \mathbf{w}_n\}$ after the T-Encoder, and entity embeddings $\{e_1, \ldots, e_m\}$, the prediction is computed as

$$p(\mathbf{e}_j|\mathbf{w}_i) = \frac{\exp\left(\mathrm{linear}(\mathbf{w}_i^o) \cdot \mathbf{e}_j\right)}{\sum_{k=1}^{m} \exp\left(\mathrm{linear}(\mathbf{w}_i^o) \cdot \mathbf{e}_k\right)} \tag{3.8}$$

The pretrained ERNIE model has improved performance on both entity-related and general tasks. It improves performance on entity typing and relation classification and also achieves new state-of-the-art results on the general GLUE benchmark [48].

3.5.2 Entity as Experts

Févry et al. [24] proposes Entity as Experts (EaE) which trains its own entity embeddings from the language modeling task. It uses hyperlinks in Wikipedia to identify entity mentions in Wikipedia pages. Every mention corresponds to an entity, and it keeps a memory to store all learned entity embeddings.[1] For every entity mention m_i, EaE first computes a pseudo-embedding h_{m_i} based on its span representation:

$$h_{m_i} = \mathbf{W_f}[x^l_{s_{m_i}} \| x^l_{t_{m_i}}] \tag{3.9}$$

where $x^l_{s_{m_i}}$ and $x^l_{t_{m_i}}$ are the start and tail representations of m_i respectively. Then it finds the k nearest neighbors in the entity memory. More formally,

$$E_{m_i} = \sum_{e_j \in \text{top}K(E,h_{m_i},k)} \alpha_j \cdot \text{EntEmbed}(e_j), \tag{3.10}$$

$$\alpha_j = \frac{\exp(\text{EntEmbed}(e_j) \cdot h_{m_i})}{\sum_{e \in \text{top}K(E,h_{m_i},k)} \exp(\text{EntEmbed}(e) \cdot h_{m_i})}, \tag{3.11}$$

where $\text{EntEmbed}()$ is the entity embedding memory, and $\text{top}K$ find the K-nearest neighbor of h_{m_i}.

EaE trains the language model on entity linking, mention detection, and language modeling tasks. Entity linking is enforced by making the pseudo embedding close to the entity embedding in memory:

$$\text{ELLoss} = \sum_{m_i} \alpha_i \cdot \mathbf{1}_{\{e_{m_i} \neq e_\emptyset\}}, \tag{3.12}$$

$$\alpha_i = \frac{\exp(\text{EntEmbed}(e_{m_i}) \cdot h_{m_i})}{\sum_{e \in E} \exp(\text{EntEmbed}(e) \cdot h_{m_i})}, \tag{3.13}$$

EaE improves the language model's performance on the LAMA probing task. It also improves the result on open-domain QA tasks including TriviaQA and WebQuestions.

---

[1] In practice, EaE only keeps embeddings for entities appearing 10 times or more.

3.5.3 Dict-BERT

Dict-BERT [29] adds definitions of rare words from the dictionary to help NLU. Here rare words are the least frequent words that comprise 10% of the total number of words. Dict-BERT proposes word-level and sentence-level auxiliary losses to enable language models to use dictionary definitions:

- At word level, the model tries to maximize the mutual information between the contextual representation of rare words in the input and in its definition. In practice, Dict-BERT uses InfoNCE [49] to approximate this mutual information.
- At sentence level, the model tries to identify the correct definition. A random portion of definitions are polluted with random definitions, and the model needs to identify if a definition is indeed relevant to the given sentence.

Together with the traditional MLM loss, Dict-BERT achieves improved performance on the general GLUE benchmark, as well as in specialized domain datasets for biomedical science, computer science, news text, and e-commerce reviews. Notably, the improvement in specialized domains are higher than general domain, likely because rare word definitions are much more critical in these specialized domains.

3.5.4 Dense Passage Retrieval

Dense Passage Retrieval (DPR, [41]) proposes the first continuous retrieval system for open-domain question answering. It uses two BERT models to encode the question and passages separately. Similarity between question and passages are computed through cosine similarity:

$$\text{sim}(p, q) = E_Q(q)^T E_P(p), \tag{3.14}$$

where E_Q and E_P are the question and passage encoders respectively.

DPR uses contrastive loss to train E_Q and E_P. For every question q_i, a ground truth passage p_i^+ contains the correct answer. DPR samples n negative passage without the correct answer: $p_{i,1}^-, \ldots, p_{i,n}^-$. Then it optimizes the negative log-likelihood to select p_i^+:

$$L(q_i, p_i^+, p_{i,1}^-, \ldots, p_{i,n}^-) = -\log\left(\frac{e^{\text{sim}(q_i, p_i^+)}}{e^{\text{sim}(q_i, p_i^+)} + \sum_{j=1}^{n} e^{\text{sim}(q_i, p_{i,j}^-)}}\right) \tag{3.15}$$

DPR experiments with random negatives or hard negatives selected by BM25, or gold passages for other questions as negatives. It also uses in-batch negative examples in the same minibatch. Results show that gold+BM25 obtains the best retrieval performance.

During inference, DPR uses FAISS [44] to accelerate the retrieval of top 5-100 passages. The passages are combined through ensemble: the model makes one prediction for every question+retrieved passage pair, and do a majority vote across passages. DPR substantially improves the state-of-the-art performance across many ODQA benchmarks.

3.5.5 REALM

Guu et al. [46] proposes a retrieval-augmented language model (REALM) pertaining, which applies the dense retrieval idea for LM pretraining. It uses similar BERT encoders for query and document as DPR; but the query here is instead an input sentence with masked tokens for language modeling. The retrieved passages are concatenated with the original input to let the model reason between them:

$$\texttt{join}(x, z) = [\text{CLS}]x[\text{SEP}]z[\text{SEP}] \tag{3.16}$$

where z is the retrieved documents for x.

One challenge for large-scale pretraining is that the index has to change over time to reflect changes in the retriever. DPR performs asynchronous index updates by using a primary *trainer* job and a secondary *index builder* job. To find harder examples for language modeling, REALM proposes salient span masking, which masks salient spans that are named entities, and mask out whole entities.

3.5.6 REINA

Wang et al. [40] proposes to retrieve from a very different source: training data. For any given test data, it retrieves the closest training examples in the training set, hoping that the knowledge from training examples can also help the test data. REINA applies to a wide range of tasks including both language understanding and generation.

REINA first builds the training data as an input-output key-value index. Given a new input, the query is used as the key to find the closest match in the index, and the corresponding values are used as the knowledge context. Different tasks use a different format as key-value pairs:

- Summarization: Documents are keys, and summaries are values.
- Language Modeling: The current chunk is the key, and the next chunk is the value.
- Machine Translation: Source text is the key, and the target text is the value.

- Question Answering: Concatenation of Q+A is treated as both key and value. Alternatively, the key can be the concatenation of Q, A, and additional knowledge.

Relevant knowledge is retrieved using BM25 discrete retriever. Then REINA appends relevant knowledge to the corresponding input and fine-tunes the language model.

REINA achieves universal improvement on both NLU and NLG tasks.

3.5.7 KEAR

Xu et al. [31] propose KEAR, which combines structured and unstructured knowledge to achieve human-level performance on the CommonsenseQA benchmark. For a given question with multiple choices, KEAR uses string matching to find relevant entities in the ConceptNet knowledge graph. Knowledge is then retrieved from three different sources:

- Knowledge Graph: KEAR takes edges between entities in the question and entities in the choices.
- Dictionary: Definition of key entities in both question and choices are selected as additional input.
- Training data: KEAR also uses the REINA method to use relevant training data as knowledge.

All the retrieved knowledge are linearized and appended as input to the language model. With additional techniques like virtual adversarial training, [50], careful selection of pretrained language model, and ensemble methods, KEAR achieves the human parity on CommonsenseQA.

3.5.8 REPLUG

The emergence of large language models (LLMs) represents a unique challenge for existing methods to incorporate external knowledge. One is the large size: LLMs often only have APIs to query, and model weights are not public. The large size also makes it prohibitive to fine-tune LLMs to adapt to new knowledge. On the other hand, LLMs store large amounts of knowledge inside their own weights, therefore useful knowledge for small models might not be still useful for LLMs.

To resolve this problem, REPLUG [51] proposes methods to train retrievers for black-box LLMs. For any input x and retrieved document D, it computes a score

$$PR(d|x) = \frac{e^{s(d,x)/\gamma}}{\sum_{d \in D'} e^{s(d,x)/\gamma}} \tag{3.17}$$

where s is the scoring function of the retriever, and γ is a hyperparameter. The black-box LM is also used as a scoring function for documents:

$$Q(d|x, y) = \frac{e^{PLM(y|d,x)/\beta}}{\sum_{d \in D'} e^{PLM(y|d,x)/\beta}} \tag{3.18}$$

where $PLM(y|d, x)$ is the probability of the correct answer with (d, x) as input, and β is a hyperparameter. REPLUG trains the retriever to minimize the KL divergence between PR and Q. REPLUG improves GPT-2 and GPT-3 models on language modeling, multitask NLU, and open domain QA tasks.

3.6   Conclusion and Future Trends

In this chapter, we cover models and methods on external knowledge for language understanding tasks. We formalize the knowledge incorporation process into three steps and introduce methods in each step. External knowledge proves useful for many NLU tasks, and both structured and unstructured knowledge can help boost NLU performance.

Many future directions exist. The emergence of large language models (LLMs) creates new challenges for knowledge integration methods: how to find useful information for an LLM, and how to efficiently incorporate knowledge into LLM inference are both critical directions. Another direction is multimodal knowledge: in addition to the textual information, multimodal data such as images and audio can also be valuable resources for knowledge retrieval.

References

1. Kwiatkowski, T., Palomaki, J., Redfield, O., Collins, M., Parikh, A., Alberti, C., Epstein, D., Polosukhin, I., Devlin, J., Lee, K., Toutanova, K., Jones, L., Kelcey, M., Chang, M.-W., Dai, A.M., Uszkoreit, J., Le, Q., Petrov, S.: Natural questions: A benchmark for question answering research. Trans. Assoc. Comput. Linguist. **7**, 452–466 (2019)
2. Joshi, M., Choi, E., Weld, D., Zettlemoyer, L.: TriviaQA: a large scale distantly supervised challenge dataset for reading comprehension. In: Proceedings of the 55th Annual Meeting of the Association for Computational Linguistics (Volume 1: Long Papers) Vancouver, Canada, July 2017, pp. 1601–1611. Association for Computational Linguistics (2017)
3. Fan, A., Jernite, Y., Perez, E., Grangier, D., Weston, J., Auli, M.: ELI5: long form question answering. In: Korhonen, A., Traum, D.R., Màrquez, L. (eds.) Proceedings of the 57th Conference of the Association for Computational Linguistics, ACL 2019, Florence, Italy, July 28- August 2, 2019, Volume 1: Long Papers, pp. 3558–3567. Association for Computational Linguistics (2019)

4. Yang, Z., Qi, P., Zhang, S., Bengio, Y., Cohen, W., Salakhutdinov, R., Manning, C.D.: HotpotQA: a dataset for diverse, explainable multi-hop question answering. In: Proceedings of the 2018 Conference on Empirical Methods in Natural Language Processing, Brussels, Belgium, October-November 2018, pp. 2369–2380. Association for Computational Linguistics (2018)

5. Berant, J., Frostig, A.C.R., Liang, P.: Semantic parsing on Freebase from question-answer pairs. In: Proceedings of the 2013 Conference on Empirical Methods in Natural Language Processing, Seattle, Washington, USA, October 2013, pp. 1533–1544. Association for Computational Linguistics (2013)

6. Talmor, A., Herzig, J., Lourie, N., Berant, J.: CommonsenseQA: A question answering challenge targeting commonsense knowledge. In: Proceedings of the 2019 Conference of the North American Chapter of the Association for Computational Linguistics: Human Language Technologies, Volume 1 (Long and Short Papers), Minneapolis, Minnesota, June 2019, pp. 4149–4158. Association for Computational Linguistics (2019)

7. Speer, R., Chin, J., Havasi, C.: Conceptnet 5.5: an open multilingual graph of general knowledge. Proceedings of the AAAI conference on artificial intelligence, vol. 31, no. 1 (2017)

8. Sap, M., Rashkin, H., Chen, D., Le Bras, R., Choi, Y.: Social IQa: commonsense reasoning about social interactions. In: Proceedings of the 2019 Conference on Empirical Methods in Natural Language Processing and the 9th International Joint Conference on Natural Language Processing (EMNLP-IJCNLP), Hong Kong, China, 2019, pp. 4463–4473. Association for Computational Linguistics (2019)

9. Bisk, Y., Zellers, R., LeBras, R., Gao, J., Choi, Y.: PIQA: reasoning about physical commonsense in natural language. In: The Thirty-Fourth AAAI Conference on Artificial Intelligence, AAAI 2020 (2020)

10. Rajpurkar, P., Zhang, J., Lopyrev, K., Liang, P.: SQuAD: 100,000+ questions for machine comprehension of text. In: Proceedings of the 2016 Conference on Empirical Methods in Natural Language Processing, Austin, Texas, November 2016, pp. 2383–2392. Association for Computational Linguistics (2016)

11. Sun, K., Yu, D., Chen, J., Yu, D., Cardie, C.: Improving machine reading comprehension with contextualized commonsense knowledge. In: Proceedings of the 60th Annual Meeting of the Association for Computational Linguistics (Volume 1: Long Papers), Dublin, Ireland, May 2022, pp. 8736–8747. Association for Computational Linguistics (2022)

12. Guo, Z., Barbosa, D.: Robust named entity disambiguation with random walks. Semantic Web 9(4), 459–479 (2018)

13. Cattaneo, A., Justus, D., Mellor, H., Orr, D., Maloberti, J., Liu, Z., Farnsworth, T., Fitzgibbon, A., Banaszewski, B., Luschi, C.: Ogb-lsc: Wikikg90mv2 Technical Report (2022)

14. Miller, A.H., Lewis, P., Bakhtin, A., Wu, Y., Petroni, F., Rocktäschel, T., Riedel, S.: Language models as knowledge bases? In: Proceedings of the 2019 Conference on Empirical Methods in Natural Language Processing (EMNLP), 2019 (2019)

15. Thorne, J., Vlachos, A., Christodoulopoulos, C., Mittal, A.: FEVER: a large-scale dataset for fact extraction and VERification. In: Proceedings of the 2018 Conference of the North American Chapter of the Association for Computational Linguistics: Human Language Technologies, Volume 1 (Long Papers), New Orleans, Louisiana, June 2018, pp. 809–819. Association for Computational Linguistics (2018)

16. Dinan, E., Roller, S., Shuster, K., Fan, A., Auli, M., Weston, J.: Wizard of wikipedia: Knowledge-powered conversational agents. In: 7th International Conference on Learning Representations (2019)

17. Levesque, H., Davis, E., Morgenstern, L.: The winograd schema challenge. In: Thirteenth International Conference on the Principles of Knowledge Representation and Reasoning (2012)

18. Zellers, R., Holtzman, A., Bisk, Y., Farhadi, A., Choi, Y.: Hellaswag: Can a machine really finish your sentence? In Proceedings of the 57th Annual Meeting of the Association for Computational Linguistics (2019)

19. Petroni, F., Piktus, A., Fan, A., Lewis, P., Yazdani, M., De Cao, N., Thorne, J., Jernite, Y., Karpukhin, V., Maillard, J., Plachouras, V., Rocktäschel, T., Riedel, S.: KILT: a benchmark

for knowledge intensive language tasks. In: Proceedings of the 2021 Conference of the North American Chapter of the Association for Computational Linguistics: Human Language Technologies, Online, June 2021, pp. 2523–2544. Association for Computational Linguistics (2021)

20. Thakur, N., Reimers, N., Rücklé, A., Srivastava, A., Gurevych, I.: BEIR: a heterogeneous benchmark for zero-shot evaluation of information retrieval models. In Thirty-fifth Conference on Neural Information Processing Systems Datasets and Benchmarks Track (Round 2) (2021)

21. Zhang, Z., Han, X., Liu, Z., Jiang, X., Sun, M., Liu, Q.: ERNIE: Enhanced language representation with informative entities. In Proceedings of the 57th Annual Meeting of the Association for Computational Linguistics, Florence, Italy, 2019, pp. 1441–1451. Association for Computational Linguistics (2019)

22. Bordes, A., Usunier, N., García-Durín, A.,Weston, J., Yakhnenko, O.: Translating embeddings for modeling multi-relational data. In: Burges, C.J.C., Bottou, L., Ghahramani, Z., Weinberger, K.Q. (eds.) Advances in Neural Information Processing Systems, vol. 26. Curran Associates, Inc., New York (2013)

23. Wang, Q., Mao, Z., Wang, B., Guo, L.: Knowledge graph embedding: a survey of approaches and applications. IEEE Trans. Knowl. Data Eng. **29**(12), 2724–2743 (2017)

24. Févry, T., Soares, L.B., FitzGerald, N., Choi, E., Kwiatkowski, T.: Entities as experts: Sparse memory access with entity supervision. In Proceedings of the 2020 Conference on Empirical Methods in Natural Language Processing (EMNLP), Online, 2020, pp. 4937–4951. Association for Computational Linguistics (2020)

25. Verga, P., Sun, H., Soares, L.B., Cohen, W: Adaptable and interpretable neural MemoryOver symbolic knowledge. In: Proceedings of the 2021 Conference of the North American Chapter of the Association for Computational Linguistics: Human Language Technologies, Online, June 2021, pp. 3678–3691. Association for Computational Linguistics (2021)

26. Hu, Z., Xu, Y., Yu, W., Wang, S., Yang, Z., Zhu, C., Chang, K.-W., Sun, Y.: Empowering language models with knowledge graph reasoning for open-domain question answering. In: Proceedings of the 2022 Conference on Empirical Methods in Natural Language Processing, Abu Dhabi, United Arab Emirates, December 2022, pp. 9562–9581. Association for Computational Linguistics (2022)

27. Wang, X., Gao, T., Zhu, Z., Liu, Z., Li, J., Tang, J.: Kepler: A unified model for knowledge embedding and pre-trained language representation. Preprint (2019). arXiv:1911.06136

28. Liu, Y., Ott, M., Goyal, N., Du, J., Joshi, M., Chen, D., Levy, O., Lewis, M., Zettlemoyer, L., Stoyanov, V.: Roberta: a robustly optimized bert pretraining approach. Preprint (2019). arXiv:1907.11692

29. Yu, W., Zhu, C., Fang, Y., Yu, D., Wang, S., Xu, Y., Zeng, M., Jiang, M.: Dict-BERT: enhancing language model pre-training with dictionary. In Findings of the Association for Computational Linguistics: ACL 2022, Dublin, Ireland, May 2022, pp. 1907–1918. Association for Computational Linguistics (2022)

30. Yu, D., Zhu, C., Yang, Y., Zeng, M.: Jaket: Joint pre-training of knowledge graph and language understanding. AAAI Conference on Artificial Intelligence (AAAI) (2022)

31. Xu, Y., Zhu, C., Wang, S., Sun, S., Cheng, H., Liu, X., Gao, J., He, P., Zeng, M., Huang, X.: Human parity on commonsenseqa: Augmenting self-attention with external attention. In: De Raedt, L. (ed.) Proceedings of the Thirty-First International Joint Conference on Artificial Intelligence, IJCAI-22. International Joint Conferences on Artificial Intelligence Organization, pp. 2762–2768. Main Track (2022)

32. Ferragina, P., Scaiella, U.: Tagme: on-the-fly annotation of short text fragments (by wikipedia entities). In: Proceedings of the 19th ACM International Conference on Information and Knowledge Management, pp. 1625–1628 (2010)

33. Xu, Y., Zhu, C., Xu, R., Liu, Y., Zeng, M., Huang, X.: Fusing context into knowledge graph for commonsense reasoning. In: ACL (2021)

34. Oguz, B., Chen, X., Karpukhin, V., Peshterliev, S., Okhonko, D., Schlichtkrull, M., Gupta, S., Mehdad, Y., Yih, S.: UniK-QA: unified representations of structured and unstructured knowledge for open-domain question answering. In: Findings of the Association for Computational Linguistics: NAACL 2022, Seattle, United States, July 2022, pp. 1535–1546. Association for Computational Linguistics (2022)

35. De Cao, N., Izacard, G., Riedel, S., Petroni, F.: Autoregressive entity retrieval. In: 9th International Conference on Learning Representations, ICLR 2021, Virtual Event, Austria, May 3–7, 2021. OpenReview.net (2021)

36. Yu, W., Zhu, C., Fang, Y., Yu, D., Wang, S., Xu, Y., Zeng, M., Jiang, M.: Dict-bert: enhancing language model pre-training with dictionary. Annual Meeting of the Association for Computational Linguistics (ACL) (2022)

37. Oguz, B., Chen, X., Karpukhin, V., Peshterliev, S., Okhonko, D., Schlichtkrull,M.S., Gupta, S., Mehdad, Y., Yih, S.: Unified open-domain question answering with structured and unstructured knowledge. In: Carpuat, M., de Marneffe, M.-C., Meza, R., Ivan, V. (eds.) Findings of the Association for Computational Linguistics: NAACL 2022, pp. 1535–1546. Association for Computational Linguistics, Seattle (2020). https://doi.org/10.18653/v1/2022.findings-naacl. 115. https://aclanthology.org/2022.findings-naacl.115

38. Robertson, S., Zaragoza, H., et al.: The probabilistic relevance framework: Bm25 and beyond. Found. Trends® Inf. Retr. 3(4), 333–389 (2009)

39. Chen, D., Fisch, A., Weston, J., Bordes, A.: Reading Wikipedia to answer open-domain questions. In: Barzilay, R., Kan, M.-Y. (eds.) Proceedings of the 55th Annual Meeting of the Association for Computational Linguistics (Volume 1: Long Papers), Vancouver, Canada, July 2017, pp. 1870–1879. Association for Computational Linguistics (2017)

40. Wang, S., Xu, Y., Fang, Y., Liu, Y., Sun, S., Xu, R., Zhu, C., Zeng, M.: Training data is more valuable than you think: a simple and effective method by retrieving from training data. In Proceedings of the 60th Annual Meeting of the Association for Computational Linguistics (Volume 1: Long Papers), pp. 3170–3179 (2022)

41. Karpukhin, V., Oguz, B., Min, S., Lewis, P., Wu, L., Edunov, S., Chen, D., Yih, W.-T.: Dense passage retrieval for open-domain question answering. In: Proceedings of the 2020 Conference on Empirical Methods in Natural Language Processing (EMNLP), Online (2020)

42. Khattab, O., Potts, C., Zaharia, M.: Relevance-guided supervision for OpenQA with ColBERT. Trans. Assoc. Comput. Linguist. 9, 929–944 (2021)

43. Weinberger, K., Dasgupta, A., Langford, J., Smola, A., Attenberg, J.: Feature hashing for large scale multitask learning. In: Proceedings of the 26th Annual International Conference on Machine Learning, pp. 1113–1120 (2009)

44. Johnson, J., Douze, M., Jégou, H.: Billion-scale similarity search with gpus. IEEE Trans. Big Data 7(3), 535–547 (2019)

45. Borgeaud, S., Mensch, A., Hoffmann, J., Cai, T., Rutherford, E., Millican, K., Van Den Driessche, G.B., Lespiau, J.-B., Damoc, B., Clark, A., et al.: Improving language models by retrieving from trillions of tokens. In: International conference on machine learning, pp. 2206–2240. PMLR (2022)

46. Guu, K., Lee, K., Tung, A., Pasupat, P., Chang, M.-W.: Realm: Retrieval-augmented language model pre-training. Preprint (2020). arXiv:2002.08909

47. Izacard, G., Grave, E.: Leveraging passage retrieval with generative models for open domain question answering. In: Merlo, P., Tiedemann, J., Tsarfaty, R. (eds.) Proceedings of the 16th Conference of the European Chapter of the Association for Computational Linguistics: Main Volume, Online, April 2021, pp. 874–880. Association for Computational Linguistics (2021)

48. Wang, A., Singh, A., Michael, J., Hill, F., Levy, O., Bowman, S.: GLUE: a multi-task benchmark and analysis platform for natural language understanding. In: Linzen, T., Chrupała, G., Alishahi, A. (eds.) Proceedings of the 2018 EMNLP Workshop BlackboxNLP: Analyzing and Interpreting Neural Networks for NLP, Brussels, Belgium, November 2018, pp. 353–355. Association for Computational Linguistics (2018)

49. van den Oord, A., Li, Y., Vinyals, O.: Representation learning with contrastive predictive coding. Preprint (2018). arXiv:1807.03748

50. Jiang, H., He, P., Chen, W., Liu, X., Gao, J., Zhao, T.: SMART: Robust and efficient fine-tuning for pre-trained natural language models through principled regularized optimization. In Jurafsky, D., Chai, J., Schluter, N., Tetreault, J. (eds.) Proceedings of the 58th Annual Meeting of the Association for Computational Linguistics, Online, July 2020, pp. 2177–2190. Association for Computational Linguistics (2020)
51. Shi, W., Min, S., Yasunaga, M., Seo, M., James, R., Lewis, M., Zettlemoyer, L., Yih, W.-T.: Replug: retrieval-augmented black-box language models. Preprint (2023). arXiv:2301.12652

Chapter 4
Knowledge-augmented Methods for Natural Language Generation

Abstract Knowledge-enhanced natural language generation (NLG) represents a significant advancement in the field of artificial intelligence, focusing on the integration of diverse knowledge sources into language models to produce more accurate, coherent, and contextually relevant text. This approach addresses several challenges inherent in traditional NLG systems, such as content hallucination, lack of coherence, grammatical inaccuracies, and limitations in handling complex calculations and low-resource languages. By incorporating external knowledge, the NLG systems can effectively mitigate these issues. These systems are particularly adept at maintaining factual accuracy, ensuring grammatical and structural integrity, and responding to current events and trends. Various NLG applications such as summarization, question answering and creative writing, have demonstrated the effectiveness of this approach, with models able to generate structured summaries, provide detailed answers with supplementary information, and create coherent content aligned with commonsense knowledge. Overall, knowledge-enhanced NLG represents a significant advancement in the field, offering solutions to longstanding challenges and setting new benchmarks for accuracy, coherence, and context-awareness in automated text generation.

Keywords Natural language generation · Knowledge-enhanced methods · Knowledge retrieval · Tool utilization · Knowledge base

4.1 Introduction

4.1.1 What is Natural Language Generation?

Natural language generation (NLG) represents a multifaceted and demanding undertaking within the field of natural language processing (NLP), where the objective is to generate coherent and contextually appropriate text from various types of input data [1]. This includes textual information, numerical data, image data, structured knowledge bases, and knowledge graphs. Researchers have developed numerous technologies for this task in a wide range of applications [2–4]. For example,

© The Author(s), under exclusive license to Springer Nature Singapore Pte Ltd. 2024   41
M. Jiang et al., *Knowledge-augmented Methods for Natural Language Processing*,
SpringerBriefs in Computer Science, https://doi.org/10.1007/978-981-97-0747-8_4

machine translation generates text in a different language based on the source text; summarization generates an abridged version of the source text to include salient information; question answering generates textual answers to given questions; dialogue system supports chatbots to communicate with humans with generated responses.

The inception of deep learning technologies has catalyzed significant progress in NLG, enabling the development of sophisticated deep neural models proficient in natural language understanding and generation [5]. Within this context, the encoder-decoder model serves as a foundational construct, transforming a given input sequence into a desired output sequence [6]. This foundation has spurred innovation in NLG technologies, leading to the introduction of diverse encoder-decoder models such as recurrent neural network (RNN) encoder-decoder [6], convolutional neural network (CNN) encoder-decoder [7], and the breakthrough Transformer encoder-decoder architecture [8]. These methods have not only contributed to the methodological richness of NLG but also enhanced the quality of generated texts.

Recent advancements in NLG models, such as T5 [9] and GPT-3 [10], with their improved scale and training strategies, have substantially broadened the horizons of the field. Strikingly, these models pre-trained on extensive web data have shown the capability to internalize a comprehensive knowledge within model parameters. In addition of encoder-encoder architecture, there are growing number of pre-trained language models adopt a decoder-only approach, reflecting the diversity and adaptability of current methodologies. These decoder-only architectures, such as those employed in models like GPT-3 [10], offer several distinct advantages, such as simplified architecture and flexibility in task handling [11]. Remarkably, these state-of-the-art models, pre-trained on vast amounts of web data, demonstrate an ability to internalize an extensive spectrum of knowledge within their parameters. This achievement resonates with the broader goals of artificial intelligence, facilitating a more profound comprehension of human language and communication patterns.

In conclusion, NLG stands as a dynamic and evolving discipline within NLP, marked by continuous advancements in modeling, methods, and applications. The successful fusion of deep learning with traditional linguistic theories has paved the way for machines to emulate human-like language proficiency. The path forward promises even more innovative approaches and applications, further blurring the lines between human communication and machine-generated language, and setting new horizons for interdisciplinary research and technological development.

4.1.2 Why Knowledge Is Needed in Language Generation?

As we venture further into the era of deep learning, remarkable progress is visible in the capabilities of natural language generation models. These advancements, however, have not been without their challenges. To transcend the existing limitations, the infusion of knowledge within NLG models is becoming an indispensable element. The reasons for this vital integration are elucidated below:

1. **Hallucination**: The propensity of language models to generate content that diverges from real-world facts or source text, known as hallucination, can lead to inconsistencies and inaccuracies [12, 13]. For example, in question answering, NLG models might generate factually incorrect answers, such as stating "Lyon" as the capital of France instead of "Paris". In summarization, NLG models could generate content that, although congruent with world knowledge, is not deducible from the source, such as asserting a viewpoint never discussed in the original text.

2. **Coherence**: Generating coherent long texts is a challenging task in NLG, often hindered by difficulties in capturing long-distance dependencies [14]. Without a proper connection to world knowledge, the generated text might lack a logical flow, resulting in a fragmented narrative. For example, in story generation or long-form question answering, NLG models may falter, providing non-coherent or inconsistent content that detracts from a seamless and engaging reader experience.

3. **Grammar and Structure**: Maintaining grammatical integrity and structural norms is crucial in specialized language generation tasks such as code generation (e.g., Python, SQL) or formatted data generation (e.g., JSON, YAML) [15–17]. Lack of domain-specific knowledge may lead to syntactical errors and non-functional outputs. For example, without proper constraints, NLG models might generate non-executable code, which compromises both usability and quality.

4. **Complex Calculation**: Certain tasks necessitate NLG models to generate content based on complex calculations or sophisticated logical reasoning [18–20]. Without proper logical knowledge, or access to tools like calculators, models might struggle with these tasks. Errors in simple arithmetic or logical proofs are illustrative of the limitations faced by models in handling computational tasks.

5. **Low Resource Languages**: Models primarily trained on widely-used languages may struggle with low-resource languages, resulting in a noticeable decline in performance [21, 22]. The limitations of fixed parameter size inherently restrict the ability to fully capture all possible world languages. For example, a model trained mainly in English might generate grammatical errors in Chinese, reflecting a lack of responsiveness to diverse linguistic contexts.

6. **Real-time Information**: Once pre-trained, language models might find it difficult to access real-time information, utilize tools, or interact with APIs [18, 23, 24]. This limitation stems from the static nature of the models after pre-training, where they are disconnected from live data and unable to engage dynamically with external systems. For example, without integration with stock market APIs, models would make mistakes when providing real-time data, like stock updates or waiting list information, thereby limiting its practical application.

In conclusion, the need for knowledge in language generation transcends mere improvement of existing capabilities; it marks a fundamental requirement for advancing the field. By embedding knowledge across various domains, models can become more precise, versatile, and aligned with human understanding, thus paving the way for more reliable and intelligent NLG systems. The integration of knowledge into NLG holds the promise to address the challenges of hallucination, coherence, grammatical integrity, complex calculations, low resource language handling, and real-time responsiveness, setting the stage for the next frontier in NLG [25].

4.1.3   What Is Knowledge-enhanced NLG?

Knowledge-augmented NLG endeavors to enhance language models by intricately weaving in a vast and diverse array of knowledge sources. These sources encompass a wide range, from factual data and everyday commonsense to intricate mathematical principles and specialized domain-specific insights. This integration enhances the model understanding of input information, ensuring that the generated text is nuanced, relevant, and reflective of the multifaceted world in which we live.

Problem 4.1 (Knowledge-enhanced Text Generation) Given a standard text generation problem where an input sequence X aims to produce an output sequence Y, the introduction of additional knowledge, denoted as K, can significantly elevate the output. Thus, knowledge-augmented text generation focuses on bolstering the generation of Y from X by weaving in the vital strands of knowledge K, taking advantage of the complex interrelations between input text, knowledge, and output text.

At its essence, a NLG system adeptly translates diverse inputs–be they text, images, or other modalities–into coherent and meaningful outputs. This transformative ability is greatly magnified when paired with external knowledge sources, culminating in outputs that are not only content-rich but also deeply informed and contextual. In addressing the mentioned challenges (in Sect. 4.1.2), a knowledge-augmented NLG system offers the following solutions and advancements:

1. **Mitigating Hallucinations**: Language models could occasionally suffer from "catastrophic forgetting" after pre-training on massive web data, leading to content hallucinations. Drawing upon external knowledge bases allows the model to cross-reference inputs and minimize deviations from factual information.
2. **Enhancing Coherence**: Long-distance dependencies pose challenges for maintaining coherence. Leveraging structured external knowledge, such as from knowledge databases, provides a logical roadmap for information. By connecting to these structured knowledge bases, the NLG model is assisted in producing outputs that are contextually logical, providing readers with a seamless narrative.
3. **Ensuring Grammatical Integrity**: Ensuring grammatical and structural correctness is paramount for many NLG tasks, such as code generation. Knowledge-augmented models can be calibrated with intricate grammar rules, patterns, and structural norms. This alignment with external knowledge ensures the generated outputs adheres to syntactical and structural conventions.

4. **Facilitating Complex Calculations**: Handling intricate calculations internally can be a challenging task for NLG models. However, integrating specialized external tools, such as calculators, allows these models to delegate complex computational tasks. This approach not only ensures computational accuracy but also enhances the operational efficiency of the NLG models.

5. **Augmenting Low-Resource Language Proficiency**: Models might underperform with low-resource languages if their primary training data was in major languages, such as English. Knowledge-augmentation can bridge this gap by introducing information, grammar rules, and cultural nuances from these lesser-known languages. This ensures that generated content is linguistically and culturally accurate, catering to a diverse audience.

6. **Integrating Real-time Information**: Pre-trained NLG models have only encoded historical knowledge into their parameters, constraining their ability to respond to real-time events. In contrast, knowledge-enhanced models can be designed to interact with live data feeds, APIs, and other contemporary information sources. This dynamic feature ensures that the content generated is not solely reliant on historical data but is also aligned with current events and emerging trends, making it both relevant and timely.

Knowledge-enhanced NLG systems have showcased proficiency in generating accurate, informative and coherent text in various applications. For examples, in summarization, knowledge graph produced a structured summary and highlight the proximity of relevant concepts, when complex events related with the same entity may span multiple sentences [26]. In question answering (QA) systems, facts stored in knowledge bases completed missing information in the question and elaborate details to facilitate answer generation [27, 28]. In story generation, using commonsense knowledge acquired from knowledge graph facilitated understanding of the storyline and better narrate following plots step by step, so each step could be reflected as a link on the knowledge graph and the whole story would be a path [29].

Broadly speaking, the incorporation of knowledge into models can be segmented into three distinct methodologies. A detailed exploration of these methodologies will be presented in Sect. 4.2. Here are brief summaries of each methodology:

1. **Knowledge as Condition:** Knowledge relevant to the input text is retrieved and merged with the input. This amalgamated content then steers the decoding process.

2. **Knowledge as Constraint:** Knowledge does not merely act as a supplement; it actively constrains the decoding to ensure global consistency and accuracy.

3. **Knowledge as Feedback:** After the initial generation, knowledge sources evaluate the output, providing feedback loops for iterative refinement.

In essence, Knowledge-enhanced NLG represents a paradigm shift in how models generate content. By integrating diverse sources of knowledge, NLG systems produce outputs that are more accurate, coherent, and contextually rich. The numerous benefits range from mitigating content hallucinations to integrating real-time information, allowing these systems to cater to varied applications. The methodologies to incorporate knowledge as a condition, constraint, or feedback offer innovative strategies to harness external information. Knowledge-augmented NLG offers the promise of a more informed, and comprehensive text generation for the future.

4.2 Incorporating Knowledge into NLG Models

4.2.1 Knowledge Acquisition

Knowledge acquisition serve as the key stage in knowledge-enhanced NLG methods. In this chapter, the knowledge acquisition paradigm goes beyond just using external knowledge as supplemental information during the encoding phase, as introduced in Chap. 3; it also retrieves and conditions the model on this knowledge, directly influencing and refining the decoding and generation process.

Generally speaking, the goal of knowledge acquisition is actively seeking and incorporating external knowledge relevant to the input. This results in a richer information that synergistically combines the input and pertinent external knowledge, setting the stage for a more informed and nuanced decoding phase. Similar to Chap. 3, multiple techniques, tailored to the nature of the knowledge sources and the specific requirements of the NLG tasks, have emerged:

4.2.1.1 Sparse/Dense Retrieval

Retrieval-based methods, either sparse retrieval or dense retrieval, form the backbone of many knowledge acquisition processes. The essence of this approach lies in a dual encoder framework which calculates the similarity between input queries and potential knowledge candidates [30–34].

1. Sparse Retrieval: Rooted in traditional information retrieval like TF-IDF [35] or BM25 [30], sparse retrieval matches input queries with potential knowledge sources based on specific keywords. Its inherent simplicity offers computational advantages. However, for intricate or layered queries, it might lack precision.
2. Dense Retrieval: Standing in contrast, dense retrieval harnesses the power of embeddings. Here, both the query and the candidate knowledge sources are embedded as dense vectors within a mutual space. This method, by tapping into contextual embeddings, frequently offers superior results, particularly when the nuances and intricacies of context are paramount [32, 36–38].

A notable advantage is the universal adaptability of retrieval methods in acquiring pertinent knowledge. These methods can be employed to garner a variety of knowledge types, showcased in different structures such as tables [39], infoboxes [40], and knowledge bases [41, 42]. The data housed in these structured formats can be serialized through rule-based approaches or neural data-to-text methods, prior to being integrated into candidate documents. This heterogeneous information, once retrieved, is represented in text format [43–45].

Besides the conversion of various knowledge formats into text, recent advancements also amalgamate different knowledge representation methodologies within a single model architecture. For instance, utilizing T5 to encode text, the ensuing outputs from the textual encoder are then channeled as inputs to a GCN model which encodes the knowledge graph [42, 46]. This process culminates in a fused representation of both text and knowledge graph, thereby creating a robust framework for knowledge retrieval and representation.

4.2.1.2 Entity Linking

Entity linking serves as a bridge between the ambiguities of natural language and the structured clarity of knowledge bases [47]. By identifying specific entities in the input text and connecting them to their respective entries in structured knowledge repositories, this method enriches the NLG model's understanding of the context and the intricate relationships between entities.

To succeed in entity linking, a meticulously trained model is crucial. Leveraging advanced techniques and datasets [48, 49], such models can detect and link entities with commendable accuracy. However, like all automated processes, it's not devoid of challenges. A predominant concern is the potential for error propagation, where inaccuracies in entity recognition and linking can cascade, affecting downstream tasks and the overall coherence of the generated content.

Yet, when entity linking works flawlessly, the benefits are manifold. For instance, once an entity is recognized and linked, the model can delve into knowledge graphs to extract associated subgraphs, bringing forth a myriad of related information that would otherwise remain untapped. Similarly, structured repositories like Wikipedia often organize content based on specific entities, e.g., titles in Wikipedia passages. Recognizing and linking to these titles can unearth comprehensive passages that offer depth and context, enhancing the NLG output's richness and relevance.

4.2.1.3 Key-Value Retrieval

In addition to information retrieval and entity linking, Key-Value Retrieval stands as a straightforward yet potent method for knowledge acquisition, especially in contexts where direct matching is paramount. Central to this method is the simple premise of aligning a specific key with its corresponding value to facilitate direct information retrieval. The essence of Key-Value Retrieval is rooted in the one-to-one correspondence between a designated key, serving as the query or identifier, and its associated value, which embodies the desired information or data. This paradigm excels in delivering a direct, unambiguous retrieval process, particularly beneficial in scenarios demanding precise information extraction devoid of semantic convolution. The operational dynamics of Key-Value Retrieval are clear-cut. Upon recognizing a specific key within the input, the system swiftly navigates to the corresponding value in the designated knowledge repository. This deterministic retrieval process is often enabled through meticulously indexed, well-structured databases or lexicons.

A prime example of Key-Value Retrieval is observed in comprehensive lexicons or dictionaries. Here, the identification of a particular term (the key) triggers the system to promptly fetch its detailed definition or related content (the value) [50, 51]. This direct matching capability extends beyond lexicons to a wide array of domains, including encyclopedic entries, technical databases, and structured knowledge bases, thereby showcasing its versatility and utility in diverse knowledge acquisition.

4.2.1.4 Tool Utilization

The incorporation of external tools, such as calculators [18], search engines [18, 52], third-party services [53], and pre-trained machine learning models [23] represents a significant stride in the realm of knowledge acquisition for NLG systems. These tools serve as extensions of the system's inherent capabilities, enabling it to perform intricate calculations, traverse the vast expanse of the web, and engage specialized services to obtain and integrate knowledge with a high degree of finesse [24, 54].

Granting language models the ability to utilize tools bestows several noteworthy advantages. The first advantage is enriched capabilities. Tools equip the system to address complex queries and tasks that would otherwise remain unattainable. Whether it's performing mathematical operations via calculators, accessing current information through search engines, or exploring domain-specific knowledge through third-party services, tools usher in a level of interaction and information retrieval that markedly expands the system's operational horizon. The second is adaptive resourcefulness. Armed with tools, the NLG system morphs into an adaptive and resourceful entity. It can interact dynamically with these tools, aligning its approach based on the nature of the query at hand, thus ensuring more accurate and informed responses. This adaptive resourcefulness is pivotal in tackling a diverse array of queries and tasks, enabling the system to customize its approach to meet the unique demands of each scenario. The third is enhanced depth and precision. The engagement of tools significantly elevates the depth and precision with which the NLG system can address queries. Tools pave the way for a deeper exploration of subject matter, facilitating the extraction of nuanced information and delivering responses with a heightened degree of accuracy and relevance. This progression not only amplifies the accuracy but also enriches the informativeness of the responses, thereby enriching the user experience.

4.2.2 Knowledge Representation

Contemporary pre-trained NLG models are predominantly founded on two dominant architectures: encoder-decoder framework [9, 55–57] and decoder-only framework [10, 53, 58–60]. From a probabilistic perspective, these generation mechanisms discern the conditional distribution over a variable length sequence (i.e., output Y) contingent on another variable length sequence (i.e., input X):

$$P(Y|X) = \prod_{t=1}^{|Y|} p(y_t|y_{<t}, X). \tag{4.1}$$

It should be noted that when $t = 1$, $y_{<t}$ indicates that no output token has been produced, meaning the model depends exclusively on the input sequence. This typically represents the state before decoding commences.

Optimization The generation trajectory is conceptualized as a sequential multi-label classification conundrum. Its direct optimization can be achieved using the negative log likelihood (NLL) loss, resulting in the formulation:

$$\mathcal{L}_{NLL}(\theta) = -\log p_\theta(Y|X) = -\sum_{t=1}^{|Y|} \log\left(p_\theta(y_t|y_{<t}, X)\right). \tag{4.2}$$

4.2.2.1 Knowledge as Condition

This approach integrates an auxiliary variable Z, extending the model as follows: $p(Y|X) = \sum_i p(Y|z_i, X)p(z_i|X)$. In a pragmatic sense, it's unfeasible to aggregate all potential knowledge. Consequently, the prevalent method is to compute the MAP estimate $\hat{Z} = \text{argmax } \hat{p}(Z)$ utilizing beam search, approximating the summation over z_i with this singular value. Knowledge, in its various forms–events, facts, or logic–demands an adept mechanism to pinpoint or source the most germane fragments pertinent to a given input X. Upon identifying or retrieving the relevant knowledge Z, it's seamlessly integrated into the model to guide the generation.

Given the vast and heterogeneous landscape of global knowledge, this methodology necessitates specialized mechanisms, such as knowledge graphs, databases, or neural memory networks, to curate and access knowledge. This process can be distilled into two pivotal steps: Firstly, knowledge retrieval, where based on a given input sequence X, the most pertinent knowledge segments Z are discerned and retrieved from an expansive knowledge reservoir using sophisticated techniques like information retrieval, entity linking, or dense vector search. Secondly, conditional generation, where the generative model assimilates both the primary input X and the extracted knowledge Z, culminating in an output that is a harmonized blend of the original input and the enriched context derived from the knowledge.

Mathematical Formulation The conditional probability can be explicated as:

$$P(Y|X, Z) = \prod_{t=1}^{|Y|} p(y_t|y_{<t}, X, Z). \tag{4.3}$$

4.2.2.2 Knowledge as Regularization

One effective technique for assimilating knowledge into NLG models is to harness it as a form of regularization. This approach sidesteps the direct conditioning of generation on specific knowledge, opting instead to train the model to produce content that aligns with established facts or logical constructs.

Three primary methods underline the principle of Knowledge as Regularization:

Constraint Decoding Constraint decoding has its roots in the broader field of structured prediction. Structured prediction models aim to predict structured outputs (e.g., sequences, trees, or graphs) based on input data. Constraint decoding, in essence, builds on this by imposing additional conditions or 'constraints' to guide the prediction towards valid outputs. This method is prevalent in code generation tasks, encompassing languages like Python, SQL, and more. In code generation, the structured output is typically a sequence of tokens representing a code snippet in a programming language, like Python or SQL. The constraints are the syntax rules and semantics of the target language. Without constraint decoding, code generation models might produce code snippets that seem semantically correct from a high-level perspective but are riddled with syntax errors, making them non-executable.

There are several methods are available for implementing constraint decoding: (1) Masking Impermissible Tokens: During the decoding phase, certain tokens are made unavailable based on the current state of the sequence. For instance, after an "if" keyword in Python, an opening parenthesis or a variable might be expected, but not another "if". (2) Tree-Structured Decoding: Some approaches represent the code as a tree, such as an Abstract Syntax Tree (AST). Decoding in this context involves generating the tree nodes in a way that adheres to the programming language's grammar. (3) Grammar-Driven Decoding: Leveraging formal grammar rules, models like Recursive Neural Networks can be trained to generate code that strictly adheres to the syntactic rules of the target language. (4) Post-process Corrections: Another method involves generating the code first and then correcting it post-hoc using rule-based systems or additional models trained specifically for error correction.

Posterior Regularization (PR) The posterior regularization (PR) framework was proposed to restrict the space of the model posterior on unlabeled data as a way to guide the model towards desired behavior [61, 62]. PR has been used as a principled framework to impose knowledge constraints on probabilistic models (including deep networks) in general [63, 64]. PR augments any regular training objective $\mathcal{L}(\theta)$ (e.g., negative log-likelihood, as in Eq.(4.2)) with a constraint term to encode relevant knowledge. Formally, denote the constraint function as $f(X, Y) \in \mathbb{R}$ such that a higher $f(X, Y)$ value indicates a better generated sequence Y that incorporates the knowledge. PR introduces an auxiliary distribution $q(Y|X)$, and imposes the constraint on q by encouraging a large expected $f(X, Y)$ value: $\mathbb{E}q[f(X, Y)]$. Meanwhile, the model p_θ is encouraged to stay close to q through a KL divergence term. The learning problem is thus a constrained optimization:

$$\max_{\theta, q} \mathcal{L}(\theta) - \mathrm{KL}(q(Y|X)||p_\theta(Y|X)) + \xi \tag{4.4}$$

$$s.t. \ \mathbb{E}q[f(X, Y)] > \xi, \tag{4.5}$$

where ξ is the slack variable. The PR framework is also related to other constraint-driven learning methods [65, 66]. We refer readers to [61] for more discussions.

Plug and Play (PPLM) Pre-trained language models leverage large amounts of unannotated data with a simple log-likelihood training objective. Controlling language generation by particular knowledge in a pre-trained model is difficult if we do not modify the model architecture to allow for external input knowledge or fine-tuning with specific data [67]. Plug and play language model (PPLM) opened up a new way to control language generation with particular knowledge during inference. At every generation step during inference, the PPLM shifts the history matrix in the direction of the sum of two gradients: one toward higher log-likelihood of the attribute a under the conditional attribute model $p(a|Y)$ and the other toward higher log-likelihood of the unmodified pre-trained generation model $p(Y|X)$ (e.g., GPT). Specifically, the attribute model $p(a|Y)$ makes gradient based updates to ΔS_t as:

$$\Delta S_t \leftarrow \Delta S_t + \frac{\nabla_{\Delta S_t} \log p(a|S_t + \Delta S_t)}{||\nabla_{\Delta S_t} \log p(a|S_t + \Delta S_t)||^\gamma}, \tag{4.6}$$

where γ is the scaling coefficient for the normalization term; ΔS_t is update of history matrix S_t (see Eq.(4.6)) and initialized as zero. The update step is repeated multiple times. Subsequently, a forward pass through the generation model is performed to obtain the updated $\widetilde{S}_{t+1}$ as $\widetilde{S}_{t+1} = \textsc{LanguageModel}((S_t + \Delta S_t), e(y_t), H)$. The perturbed $\widetilde{S}_{t+1}$ is then used to generate a new logit vector. PPLMs is efficient and flexible to combine differentiable attribute models to steer text generation [68].

This regularization strategy proves particularly invaluable when faced with uncertainties regarding the relevance or accuracy of retrieved knowledge. Using knowledge as a malleable constraint during training, models can adeptly navigate between adhering to factual knowledge and crafting fluent, coherent prose. For instance, knowledge-driven regularization might penalize models for generating outputs that negate established facts or might incentivize outputs that harmoniously align with a designated knowledge graph.

4.2.2.3 Knowledge as Implicit Bias

Harnessing the true potential of language models often requires going beyond direct injection of external knowledge. One promising avenue is training models on datasets inherently imbued with the desired knowledge. While the general pre-training phase can introduce a model to a vast spectrum of information, there is a looming risk: catastrophic forgetting, wherein the model loses previously acquired knowledge in the process of learning new information.

To mitigate this, researchers have turned to specific strategies, one of which is the structured training on commonsense-rich corpora. Such sources, like those cited in [69–71], are not just any random datasets; they are carefully curated collections of information that reflect human-like common sense. When chosen and integrated astutely, models trained on these datasets can remarkably surpass their conventionally pre-trained counterparts in performance, as evidenced by recent findings [71]. However, the realm of knowledge embedding doesn't halt at commonsense corpora. There's a burgeoning interest in domain-specific corpora, which cater to specialized fields or niches. These corpora serve as a goldmine for language models, allowing them to delve deeper into specialized knowledge areas, thus attaining a level of expertise comparable to domain experts. One of the towering exemplars of this approach is the GPT-3 model. Its design philosophy revolves around immersing the model in an extensive web of data, enabling it to sift through and implicitly acquire diverse knowledge from its vast training corpus. This method, no doubt, exhibits unparalleled efficacy across a wide range of tasks, making the model a jack-of-all-trades.

However, as with all methodologies, there are inherent limitations. The implicit knowledge acquisition technique is bounded by the timeliness and accuracy of the training data. Given that GPT-3's knowledge is constrained to its last update, the model might not always reflect the most current or nuanced understanding of specific topics. This poses a challenge for tasks that require up-to-date information or a deep, contemporary understanding of evolving subjects.

4.2.2.4 Knowledge as Feedback

Utilizing feedback loops to embed knowledge within natural language generation models presents a dynamic methodology that perpetually refines and updates model responses. This strategy is grounded in the understanding that knowledge is not static, but evolves alongside time and context. Leveraging feedback as a mechanism to enhance knowledge capitalizes on real-world interactions and insights from users or domain experts, along with automated techniques, thereby enabling the language model to generate more precise and contextually relevant outputs.

User-Driven Feedback By allowing users to provide feedback on the helpfulness and harmlessness, the NLG model can learn and adapt to both specific user preferences and overarching content quality standards [72]. This feedback can take the form of ratings, textual corrections, or more structured annotations. Over time, consistent user feedback enables the model to capture nuances of specific domains, rectify biases, and refine knowledge representations. Besides, collaboration with domain experts provides a richer source of feedback [73]. For instance, in a medical NLG application, feedback from clinicians can ensure generated content adheres to the latest medical guidelines and standards. These expert annotations can be periodic and serve as checkpoints to validate the model's adherence to specific domain standards.

Automatic Feedback Given the resource-intensive nature of collecting human feedback, numerous studies have delved into the utilization of automated feedback to lessen the necessity for human intervention. Automatic feedback can be directed by external metrics [74, 75], which assist in optimizing model parameters based on external evaluation measures, or be guided by external knowledge sources or tools [76, 77]. Additionally, the realm of learning from AI-generated feedback also presents a promising avenue [78]. This method not only alleviates the demand on human resources but also introduces a level of continuous, real-time model refinement. Through automated monitoring and feedback mechanisms, the model's performance can be perpetually enhanced, ensuring its alignment with evolving domain-specific standards and user expectations. This approach shows the potential of intertwining automated and user-driven feedback, enabling a more robust and adaptive NLG system.

To learn with the feedback, there are two main methods, **direct optimization** and **reinforcement learning**.

Direct Optimization is a straightforward strategy involves fine-tuning the model on outputs that receive positive feedback. For instance, Sparrow [79] fine-tunes LLMs on collected dialogues rated as preferred and rule-compliant (correctness, harmfulness, and helpfulness) by human evaluators. However, solely using positive data for fine-tuning may hinder the model's capacity to identify and correct negative attributes or errors. To mitigate this, Chain-of-Hindsight [80] fine-tunes the LLM on model outputs paired with both positive and negative feedback.

Reward Modeling and RLHF Employing human feedback directly to rectify model behavior may not always be practical due to the labor-intensive and time-consuming nature of collecting human feedback. A more efficient alternative is to develop a reward model that mimics human feedback. Once trained, this reward model can provide consistent, real-time feedback for every model output, thus bypassing the need for continual human involvement. The integration of RL principles with NLG enables the optimization of content generation based on feedback signals. In this setup, feedback serves as the reward signal, with an RL agent trained to maximize these rewards over time, resulting in improved content generation. Through a feedback-driven reward system, the model progressively refines its generation process, aligning more closely with desired outcomes and human preferences [58].

4.3   Representative Methods

4.3.1   NLG Methods Enhanced by Text Retrieval

Grounded knowledge in text entails the integration of pertinent supplementary information in a given input text, transcending the boundaries of information contained within the pre-training corpora. This augmentation is derived from extensive online text repositories, with platforms like Wikipedia furnishing comprehensive elucidations and contextual understanding. Retrieval-augmented generation (RAG) emerges as a novel learning paradigm utilized to fortify the input of language models with external information. Highlighted in studies such as those by Guu et al. [81], Lewis et al. [82], Izacard and Grave [83], Izacard et al. [84], Borgeaud et al. [85], and Yu et al. [86], RAG ingeniously melds traditional information retrieval techniques with pre-trained language models, exhibiting stellar performance across various knowledge-centric NLP tasks. More precisely, this enriched paradigm can be depicted as: $Y = f(X, Z)$, where Z symbolizes relevant data points procured from either the initial training set or external datasets. The core notion here is that if Z bears significant resemblance to the input X, it could potentially amplify the quality of the generated output. The elegance of RAG's concept underscores its efficacy: it principally comprises two components: a retrieval model and a reader model. The role of retrieval model is to unearth relevant information from a vast array of documents based on the input query through semantic search, thereby anchoring the reader models to the most precise, current information and granting users a glimpse into the generative mechanics of language models.

In summary, RAG unveils itself as a robust framework adept at interweaving external, real-time information retrieval with the generative capabilities of language models, thus enriching the textual output with grounded, contextual knowledge.

4.3.1.1   REALM

Diving into the specifics, the REALM, short for retrieval-augmented language model [81], stands out for its innovative approach to augment language model training with a specialized knowledge retrieval mechanism. The core principle of REALM is to optimize the retriever using an unsupervised training loss. This means a retrieval that enhances the language model's perplexity should be positively reinforced, while irrelevant retrievals are discouraged. As an illustration, if the task is to complete a sentence like "the ___ at the top of the pyramid", the retriever should be incentivized to reference a text such as "The pyramidion allows for reduced material at the pyramid's summit" (the example is extracted from the paper [81]). This optimization is managed by structuring the retrieve-then-predict method as a latent variable within the language model, focusing on the overall likelihood.

In both pre-training and fine-tuning stage, REALM uses an input text sequence X to predict potential output Y. REALM's methodology breaks down $p(Y|X)$ into two stages: retrieval followed by prediction. Given an input X, documents Z from a knowledge bank are retrieved, modeled as a sample from $p(Z|X)$. Post retrieval, the prediction is conditioned on both X and Z, represented as $p(Y|Z, X)$. For a comprehensive likelihood assessment, Z is treated as a latent factor, considering all plausible document references. During pre-training, the goal is masked language modeling where X_{masked} denotes a sentence with certain masked tokens, and the model's task is to ascertain the value of these tokens Y_{masked}. During the fine-tuning phase, the objective shifts to the open-domain extractive question answering task [31], where X_{query} represents a question with Y_{answer} being its answer.

Knowledge Retriever   The retriever employs a dense inner product model to determine the probability $p(Z|X)$:

$$p(Z|X) = \frac{\exp f(X, Z)}{\sum_{Z'} \exp f(X, Z')} \tag{4.7}$$

The relevance score $f(X, Z)$ between X and Z is given by the inner product of their vector embeddings transformed using BERT-style Transformers [87] and then subjected to a linear projection to lower their dimensionality.

Knowledge-Augmented Encoder   Given input X and a fetched document Z, this encoder defines $p(Y|Z, X)$. The encoder merges X and Z using a Transformer, allowing for rich cross-attention before deducing Y. For pre-training, the goal is to predict the original value of each [MASK] token in X, i.e., masked language modeling:

$$p(Y|Z, X) = \prod_{j=1}^{J_X} p(y_j|Z, X) \tag{4.8}$$

where J_X is the sum of [MASK] tokens in X. The likelihood $p(y_j|Z, X)$ is directly proportional to the exponentiation of the inner product of the Transformer output for the jth masked token and a trained word vector for token y_j. For downstream fine-tuning (i.e., open-domain question answering), the encoder presumes the response Y exists as a span in some document Z. The probability is given by:

$$p(Y|Z, X) \propto \sum_{s \in S(Z,Y)} \exp \text{MLP} [\mathbf{h}_{START}(s); \mathbf{h}_{END}(s)] \tag{4.9}$$

where $\mathbf{h}_{START}(s)$ and $\mathbf{h}_{END}(s)$ denote the Transformer output vectors for the starting and ending tokens of span s respectively.

4.3.1.2   RAG

While the REALM [81] has showcased the combination of masked language models [87] with a differentiable retriever, their focus was primarily on open-domain extractive question answering, which means the model can only extract text spans from the input text as output. Thus, the REALM cannot be used for generation tasks. To address the issue, RAG, short for retrieval-augmented generation [82] innovatively integrates the retrieval module into the sequence-to-sequence generation models [8]. The RAG used a pre-trained transformer—BART [88] as generator, with a non-parametric memory that manifests as a dense vector index of Wikipedia, accessible through a pre-trained neural retriever. This amalgamation is structured into a unified probabilistic model that undergoes comprehensive end-to-end training.

Knowledge-Augmented Generator   Central to RAG are two model variations: RAG-Sequence and RAG-Token.

1. RAG-Sequence: This avatar consistently engages the same retrieved document for each resultant token. It's sequence-to-sequence probability is articulated as:

$$p_{\text{RAG-Sequence}}(Y|X) \approx \sum_{Z \in \text{top-k}(p(\cdot|X))} p_\eta(Z|X) \prod_{i=1}^{N} p_\theta(y_i|X, Z, y_{1:i-1})$$

2. RAG-Token: This manifestation offers adaptability by potentially engaging divergent latent documents for individual target. Its probability is expounded as:

$$p_{\text{RAG-Token}}(Y|X) \approx \prod_{i=1}^{N} \sum_{Z \in \text{top-k}(p(\cdot|X))} p_\eta(Z|X) p_\theta(y_i|X, Z, y_{1:i-1})$$

4.3.1.3   FiD

REALM and RAG employ a method wherein retrieved documents are sequentially concatenated—which would swiftly reach to the input context length limitation, thereby narrowing the number of input documents. In addition to this length limitation, recent research also illustrates a notable decline in model performance as the input context increases [83]. Concurrently, the "lost in the middle" issue [89] underscores a performance apex when pertinent information is positioned at the beginning or end of the input context. However, there is a significant degradation in performance when the model needs to access relevant information embedded within extended contexts.

Fusion-in-Decoder (FiD) [83] presents a novel approach to tackle this issue by concatenating the input question with each retrieved documents separately. This concatenation is then processed independently by the encoder (T5 [9] in the experiments). The process culminates with the decoder executing attention over the collective representations of all retrieved documents. A standout feature of this model is its ability to amalgamate evidence solely within the decoder, earning it the moniker "fusion-in-decoder". Specifically, by handling documents independently within the encoder, the model demonstrates an ability to scale efficiently to a large number of contexts. A noteworthy repercussion of this architecture is that the model's computational demand increases linearly with the rise in the number of passages, as opposed to a quadratic increase in previous methods such as REALM and RAG.

4.3.1.4   RETRO

Previously introduced retrieval-augmented language models have primarily employed smaller transformers like BART [88] and T5 [9], each encompassing around 100 million parameters. They are paired with databases of a modest size, which are up to billions of tokens, which constrains the models' ability to handle expansive data efficiently and limits their capacity to cover a broader spectrum of knowledge. The retrieval-enhanced transformer (RETRO) demonstrates a significant stride in augmenting auto-regressive language models by conditioning on document chunks extracted from a much larger corpus based on local similarity with preceding tokens. With a database of *2 trillion tokens*, RETRO showcases comparable performance to GPT-3 [10] while employing 25 times fewer parameters, displaying efficiency in parameter utilization.

Compared to earlier retrieval-augmented language models which predominantly rely on smaller transformers and smaller databases, RETRO pioneers in scaling the retrieval database to trillions of tokens for large parametric language models. This scaling showcases a consistent performance gain which is comparable to a tenfold increase in the parametric model size. Moreover, the performance gain rises with the size of the retrieval database and the number of retrieved neighbors up to a point, post which it starts to degrade possibly due to reduced quality. The architecture of RETRO amalgamates a frozen Bert retriever, a differentiable encoder, and a chunked cross-attention mechanism to make token predictions based on a significantly larger volume of data compared to what is typically utilized during training. This feature signifies a move towards efficient data utilization without a proportionate increase in model size or computational demand. Unlike prior models that are trained from scratch, RETRO showcases the capability to rapidly retrofit pre-existing transformers with retrieval functionalities and still achieve commendable performance. This flexibility could expedite the adaptation of retrieval-augmented capabilities in existing models.

4.3.1.5   REPLUG

Previous methods necessitate access to the internal representations of the language model to train the model or index the datastore. This requirement restricts the applicability of these frameworks, especially with very large language models, such as ChatGPT [58], where internal representations are hard to be modified. RePLUG introduces a novel idea which treats the language model as a black box and augments it with a tunable retrieval model. This stands apart from prior retrieval-augmented langauge models which typically necessitate special cross-attention mechanisms to encode the retrieved text, or require access to internal model representations. RePLUG's framework is simplistic and flexible, as it merely prepends retrieved documents to the input for the frozen black-box language models.

One significant advantage of RePLUG is its tuneable retrieval model which, when fine-tuned, significantly improves the performance of large language models. This is achieved without having to delve into the internal representations of the language models or modifying the model parameters. Instead, RePLUG introduces a plug-and-play module that retrieves relevant documents from an external corpus and prepends them to the input context before feeding it to the black-box language model. This design makes RePLUG highly adaptable to large language models, even those accessed via APIs (e.g., ChatGPT and GPT-4) where internal representations are not exposed and fine-tuning is not supported.

References

1. Garbacea, C., Mei, Q.: Neural language generation: formulation, methods, and evaluation. Preprint (2020). arXiv:2007.15780
2. Gatt, A., Krahmer, E.: Survey of the state of the art in natural language generation: Core tasks, applications and evaluation. J. Artif. Intell. Res. (2018)
3. Wang, H., Guo, B., Wu, W., Yu, Z.: Towards information-rich, logical text generation with knowledge-enhanced neural models. Preprint (2020). arXiv:2003.00814
4. Iqbal, T., Qureshi, S.: The survey: text generation models in deep learning. J. King Saud Univ. Comput. Inf. Sci. (2020)
5. LeCun, Y., Bengio, Y., Hinton, G.: Deep learning. In: Nature. Nature Publishing Group, Berlin (2015)
6. Sutskever, I., Vinyals, O., Le, Q.V.: Sequence to sequence learning with neural networks. In: Advances in Neural Information Processing Systems (NeurIPS) (2014)
7. Gehring, J., Auli, M., Grangier, D., Yarats, D., Dauphin, Y.N.: Convolutional sequence to sequence learning. In: International Conference on Machine Learning (ICML) (2017)
8. Vaswani, A., Shazeer, N., Parmar, N., Uszkoreit, J., Jones, L., Gomez, A.N., Kaiser, Ł., Polosukhin, I.: Attention is all you need. In: Advances in Neural Information Processing Systems (NeurIPS) (2017)
9. Raffel, C., Shazeer, N., Roberts, A., Lee, K., Narang, S., Matena, M., Zhou, Y., Li, W., Liu, P.J.: Exploring the limits of transfer learning with a unified text-to-text transformer. J. Mach. Learn. Res. (2020)

10. Brown, T., Mann, B., Ryder, N., Subbiah, M., Kaplan, J.D., Dhariwal, P., Neelakantan, A., Shyam, P., Sastry, G., Askell, A., et al.: Language models are few-shot learners. Adv. Neural Inf. Process. Syst. **33**, 1877–1901 (2020)
11. Zhao, W.X., Zhou, K., Li, J., Tang, T., Wang, X., Hou, Y., Min, Y., Zhang, B., Zhang, J., Dong, Z., et al.: A survey of large language models. Preprint (2023). arXiv:2303.18223
12. Ji, Z., Lee, N., Frieske, R., Yu, T., Su, D., Xu, Y., Ishii, E., Bang, Y.J., Madotto, A., Fung, P.: Survey of hallucination in natural language generation. ACM Comput. Surv. **55**(12), 1–38 (2023)
13. Zhang, Y., Li, Y., Cui, L., Cai, D., Liu, L., Fu, T., Huang, X., Zhao, E., Zhang, Y., Chen, Y., et al.: Siren's song in the ai ocean: A survey on hallucination in large language models. Preprint (2023). arXiv:2309.01219
14. Min, S., Krishna, K., Lyu, X., Lewis, M., Yih, W.-T., Koh, P.W., Iyyer, M., Zettlemoyer, L., Hajishirzi, H.: Factscore: Fine-grained atomic evaluation of factual precision in long form text generation. Preprint (2023). arXiv:2305.14251
15. Feng, Z., Guo, D., Tang, D., Duan, N., Feng, X., Gong, M., Shou, L., Qin, B., Liu, T., Jiang, D., et al.: Codebert: A pre-trained model for programming and natural languages. In Findings of the Association for Computational Linguistics: EMNLP 2020, pp. 1536–1547 (2020)
16. De Cao, N., Izacard, G., Riedel, S., Petroni, F.: Autoregressive entity retrieval. In: International Conference on Learning Representations (2020)
17. Li, R., Allal, L.B., Zi, Y., Muennighoff, N., Kocetkov, D., Mou, C., Marone, M., Akiki, C., Li, J., Chim, J., et al.: Starcoder: may the source be with you! Preprint (2023). arXiv:2305.06161
18. Schick, T., Dwivedi-Yu, J., Dessì, R., Raileanu, R., Lomeli, M., Zettlemoyer, L., Cancedda, N., Scialom, T.: Toolformer: Language models can teach themselves to use tools. Preprint (2023). arXiv:2302.04761
19. Chen, W., Ma, X., Wang, X., Cohen, W.W.: Program of thoughts prompting: disentangling computation from reasoning for numerical reasoning tasks. Preprint (2022). arXiv:2211.12588
20. Lu, P., Peng, B., Cheng, H., Galley, M., Chang, K.-W., Wu, Y.N., Zhu, S.-C., Gao, J.: Chameleon: plug-and-play compositional reasoning with large language models. Preprint (2023). arXiv:2304.09842
21. Kim, B., Kim, J., Ko, Y., Seo, J.: Commonsense knowledge augmentation for low-resource languages via adversarial learning. In: Proceedings of the AAAI Conference on Artificial Intelligence, vol. 35, pp. 6393–6401 (2021)
22. Gupta, S., Matsubara, Y., Chadha, A., Moschitti, A.: Cross-lingual knowledge distillation for answer sentence selection in low-resource languages. Preprint (2023). arXiv:2305.16302
23. Shen, Y., Song, K., Tan, X., Li, D., Lu, W., Zhuang, Y.: Hugginggpt: solving ai tasks with chatgpt and its friends in huggingface. Preprint (2023). arXiv:2303.17580
24. Qin, Y., Liang, S., Ye, Y., Zhu, K., Yan, L., Lu, Y., Lin, Y., Cong, X., Tang, X., Qian, B. et al.: Toolllm: Facilitating large language models to master 16000+ real-world apis. Preprint (2023). arXiv:2307.16789
25. Kaddour, J., Harris, J., Mozes, M., Bradley, H., Raileanu, R., McHardy, R.: Challenges and applications of large language models. Preprint (2023). arXiv:2307.10169
26. Huang, L., Wu, L., Wang, L.: Knowledge graph-augmented abstractive summarization with semantic-driven cloze reward. In: Annual Meeting of the Association for Computational Linguistics (ACL) (2020)
27. He, S., Liu, C., Liu, K., Zhao, J.: Generating natural answers by incorporating copying and retrieving mechanisms in sequence-to-sequence learning. In: Annual Meeting of the Association for Computational Linguistics (ACL) (2017)
28. Fan, A., Gardent, C., Braud, C., Bordes, A.: Using local knowledge graph construction to scale seq2seq models to multi-document inputs. In: Conference on Empirical Methods in Natural Language Processing and International Joint Conference on Natural Language Processing (EMNLP-IJCNLP) (2019)
29. Guan, J., Wang, Y., Huang, M.: Story ending generation with incremental encoding and commonsense knowledge. In: AAAI Conference on Artificial Intelligence (AAAI) (2019)

30. Robertson, S., Zaragoza, H., et al.: The probabilistic relevance framework: Bm25 and beyond. Found. Trends® Inf. Retr. **3**(4), 333–389 (2009)
31. Chen, D., Fisch, A., Weston, J., Bordes, A.: Reading wikipedia to answer open-domain questions. In: Proceedings of the 55th Annual Meeting of the Association for Computational Linguistics (Volume 1: Long Papers), pp. 1870–1879 (2017)
32. Karpukhin, V., Oguz, B., Min, S., Lewis, P., Wu, L., Edunov, S., Chen, D., Yih, W.-T.: Dense passage retrieval for open-domain question answering. In: Proceedings of the 2020 Conference on Empirical Methods in Natural Language Processing (EMNLP), pp. 6769–6781 (2020)
33. Guo, J., Cai, Y., Fan, Y., Sun, F., Zhang, R., Cheng, X.: Semantic models for the first-stage retrieval: a comprehensive review. ACM Trans. Inf. Syst. **40**(4), 1–42 (2022)
34. Zhao, W.X., Liu, J., Ren, R., Wen, J.-R.: Dense text retrieval based on pretrained language models: a survey. Preprint (2022). arXiv:2211.14876
35. Ramos, J., et al.: Using tf-idf to determine word relevance in document queries. In: Proceedings of the First Instructional Conference on Machine Learning, vol. 242, pp. 29–48. Citeseer (2003)
36. Qu, Y., Ding, Y., Liu, J., Liu, K., Ren, R., Zhao, W.X., Dong, D., Wu, H., Wang, H.: Rocketqa: an optimized training approach to dense passage retrieval for open-domain question answering. In: Proceedings of the 2021 Conference of the North American Chapter of the Association for Computational Linguistics: Human Language Technologies, pp. 5835–5847 (2021)
37. Khattab, O., Zaharia, M.: Colbert: Efficient and effective passage search via contextualized late interaction over bert. In: Proceedings of the 43rd International ACM SIGIR Conference on Research and Development in Information Retrieval, pp. 39–48 (2020)
38. Santhanam, K., Khattab, O., Saad-Falcon, J., Potts, C., Zaharia, M.: Colbertv2: Effective and efficient retrieval via lightweight late interaction. In: Proceedings of the 2022 Conference of the North American Chapter of the Association for Computational Linguistics: Human Language Technologies, pp. 3715–3734 (2022)
39. Chen, W., Chang, M.-W., Schlinger, E., Wang, W.Y., Cohen, W.W.: Open question answering over tables and text. In: International Conference on Learning Representations (2020)
40. Wu, S., Wang, M., Zhang, D., Zhou, Y., Li, Y., Wu, Z.: Knowledge-aware dialogue generation via hierarchical infobox accessing and infobox-dialogue interaction graph network. In: IJCAI, pp. 3964–3970 (2021)
41. Yu, D., Zhu, C., Yang, Y., Zeng, M.: Jaket: joint pre-training of knowledge graph and language understanding. In: Proceedings of the AAAI Conference on Artificial Intelligence, vol. 36, no. 10, pp. 11630–11638 (2022)
42. Yu, D., Zhu, C., Fang, Y., Yu, W., Wang, S., Xu, Y., Ren, X., Yang, Y., Zeng, M.: Kg-fid: Infusing knowledge graph in fusion-in-decoder for open-domain question answering. Proceedings of the Annual Meeting of the Association for Computational Linguistics (ACL) (2022)
43. Oguz, B., Chen, X., Karpukhin, V., Peshterliev, S., Okhonko, D., Schlichtkrull, M., Gupta, S., Mehdad, Y., Yih, S.: Unik-qa: unified representations of structured and unstructured knowledge for open-domain question answering. In: Findings of the Association for Computational Linguistics: NAACL 2022, pp. 1535–1546 (2022)
44. Ma, K., Cheng, H., Liu, X., Nyberg, E., Gao, J.: Open domain question answering over virtual documents: a unified approach for data and text. Proceedings of the Annual Meeting of the Association for Computational Linguistics (ACL) (2022)
45. Xie, T., Wu, C.H., Shi, P., Zhong, R., Scholak, T., Yasunaga, M., Wu, C.-S., Zhong, M., Yin, P., Wang, S.I., et al.: Unifiedskg: unifying and multi-tasking structured knowledge grounding with text-to-text language models. In: Proceedings of the 2022 Conference on Empirical Methods in Natural Language Processing, pp. 602–631 (2022)
46. Ju, M., Yu, W., Zhao, T., Zhang, C., Ye, Y.: Grape: knowledge graph enhanced passage reader for open-domain question answering. In: Findings of the Association for Computational Linguistics: EMNLP 2022, pp. 169–181 (2022)
47. Sevgili, Ö., Shelmanov, A., Arkhipov, M., Panchenko, A., Biemann, C.: Neural entity linking: a survey of models based on deep learning. Semantic Web **13**(3), 527–570 (2022)

48. Wu, Z., Pan, S., Chen, F., Long, G., Zhang, C., Philip, S.Y.: A comprehensive survey on graph neural networks. In: IEEE Transactions on Neural Networks and Learning Systems (TNNLS) (2020)

49. De Cao, N., Izacard, G., Riedel, S., Petroni, F.: Autoregressive entity retrieval. In: International Conference on Learning Representations (2021)

50. Yu, W., Zhu, C., Fang, Y., Yu, D., Wang, S., Xu, Y., Zeng, M., Jiang, M.: Dict-bert: Enhancing language model pre-training with dictionary. Proceedings of the Annual Meeting of the Association for Computational Linguistics (ACL) (2022)

51. Yu, W., Zhu, C., Fang, Y., Yu, D., Wang, S., Xu, Y., Zeng, M., Jiang, M.: Dict-BERT: enhancing language model pre-training with dictionary. In: Findings of the Association for Computational Linguistics: ACL 2022, pp. 1907–1918 (2022)

52. Lazaridou, A., Gribovskaya, E., Stokowiec, W., Grigorev, N.: Internet-augmented language models through few-shot prompting for open-domain question answering. Preprint (2022). arXiv:2203.05115

53. OpenAI: Gpt-4 Technical Report. Preprint (2023). arXiv:2303.08774

54. Qin, Y., Hu, S., Lin, Y., Chen, W., Ding, N., Cui, G., Zeng, Z., Huang, Y., Xiao, C., Han, C., et al.: Tool learning with foundation models. Preprint (2023). arXiv:2304.08354

55. Lewis, M., Liu, Y., Goyal, N., Ghazvininejad, M., Mohamed, A., Levy, O., Stoyanov, V., Zettlemoyer, L.: BART: denoising sequence-to-sequence pre-training for natural language generation, translation, and comprehension. In: Annual Meeting of the Association for Computational Linguistics (ACL) (2020)

56. Tay, Y., Dehghani, M., Tran, V.Q., Garcia, X., Wei, J., Wang, X., Chung, H.W., Bahri, D., Schuster, T., Zheng, S., et al.: Ul2: unifying language learning paradigms. In: The Eleventh International Conference on Learning Representations (2023)

57. Chung, H.W., Hou, L., Longpre, S., Zoph, B., Tay, Y., Fedus, W., Li, E., Wang, X., Dehghani, M., Brahma, S., et al.: Scaling instruction-finetuned language models. Preprint (2022). arXiv:2210.11416

58. Ouyang, L., Wu, J., Jiang, X., Almeida, D., Wainwright, C.L., Mishkin, P., Zhang, C., Agarwal, S., Slama, K., Ray, A., et al.: Training language models to follow instructions with human feedback. Preprint (2022). arXiv:2203.02155

59. Touvron, H., Lavril, T., Izacard, G., Martinet, X., Lachaux, M.-A., Lacroix, T., Rozière, B., Goyal, N., Hambro, E., Azhar, F., et al.: Llama: open and efficient foundation language models. Preprint (2023). arXiv:2302.13971

60. Le Scao, T., Fan, A., Akiki, C., Pavlick, E., Ilić, S., Hesslow, D., Castagné, R., Luccioni, A.S., Yvon, F., Gallé, M., et al.: Bloom: a 176b-parameter open-access multilingual language model. Preprint (2022). arXiv:2211.05100

61. Ganchev, K., Gillenwater, J., Taskar, B., et al.: Posterior regularization for structured latent variable models. In: J. Mach. Learn. Res. (2010)

62. Zhu, J., Chen, N., Xing, E.P.: Bayesian inference with posterior regularization and applications to infinite latent svms. In: J. Mach. Learn. Res. (2014)

63. Hu, Z., Yang, Z., Salakhutdinov, R., Liang, X., Qin, L., Dong, H., Xing, E.: Deep generative models with learnable knowledge constraints. In: Advances in Neural Information Processing Systems (2018)

64. Zhang, J., Liu, Y., Luan, H., Xu, J., Sun, M.: Prior knowledge integration for neural machine translation using posterior regularization. In: Annual Meeting of the Association for Computational Linguistics (ACL) (2017)

65. Chang, M.-W., Ratinov, L., Roth, D.: Guiding semi-supervision with constraint-driven learning. In: Annual Meeting of the Association of Computational Linguistics (ACL) (2007)

66. Mann, G.S., McCallum, A.: Simple, robust, scalable semi-supervised learning via expectation regularization. In: International Conference on Machine Learning (ICML) (2007)

67. Dathathri, S., Madotto, A., Lan, J., Hung, J., Frank, E., Molino, P., Yosinski, J., Liu, R.: Plug and play language models: a simple approach to controlled text generation. In: International Conference for Learning Representation (ICLR) (2020)

68. Qin, L., Shwartz, V., West, P., Bhagavatula, C., Hwang, J., Le Bras, R., Bosselut, A., Choi, Y.: Backpropagation-based decoding for unsupervised counterfactual and abductive reasoning. In: Conference on Empirical Methods in Natural Language Processing (EMNLP) (2020)
69. Bosselut, A., Rashkin, H., Sap, M., Malaviya, C., Celikyilmaz, A., Choi, Y.: Comet: commonsense transformers for automatic knowledge graph construction. In: Proceedings of the 57th Annual Meeting of the Association for Computational Linguistics (ACL) (2019)
70. Lourie, N., Le Bras, R., Bhagavatula, C., Choi, Y.: Unicorn on rainbow: a universal commonsense reasoning model on a new multitask benchmark. In: Proceedings of the AAAI Conference on Artificial Intelligence (2021)
71. Zhou, W., Lee, D. H., Selvam, R.K., Lee, S., Ren, X.: Pre-training text-to-text transformers for concept-centric common sense. In: International Conference on Learning Representations (2020)
72. Griffith, S., Subramanian, K., Scholz, J., Isbell, C.L., Thomaz, A.L.: Policy shaping: integrating human feedback with reinforcement learning. Adv. Neural Inf. Process. Syst. **26** (2013).
73. Liu, Z., Wu, Z., Hu, M., Zhao, B., Zhao, L., Zhang, T., Dai, H., Chen, Z., Shen, Y., Li, S., et al.: Pharmacygpt: the ai pharmacist. Preprint (2023). arXiv:2307.10432
74. Jung, J., Qin, L., Welleck, S., Brahman, F., Bhagavatula, C., Le Bras, R., Choi, Y.: Maieutic prompting: logically consistent reasoning with recursive explanations. In: Proceedings of the 2022 Conference on Empirical Methods in Natural Language Processing, pp. 1266–1279 (2022)
75. Welleck, S., Lu, X., West, P., Brahman, F., Shen, T., Khashabi, D., Choi, Y.: Generating sequences by learning to self-correct. In: The Eleventh International Conference on Learning Representations (2023)
76. Gou, Z., Shao, Z., Gong, Y., Shen, Y., Yang, Y., Duan, N., Chen, W.: Critic: large language models can self-correct with tool-interactive critiquing. Preprint (2023)
77. Yu, W., Zhang, Z., Liang, Z., Jiang, M., Sabharwal, A: Improving language models via plug-and-play retrieval feedback. Preprint (2023). arXiv:2305.14002
78. Lee, H., Phatale, S., Mansoor, H., Lu, K., Mesnard, T., Bishop, C., Carbune, V., Rastogi, A.: Rlaif: scaling reinforcement learning from human feedback with ai feedback. Preprint (2023). arXiv:2309.00267
79. Glaese, A., McAleese, N., Trkebacz, M., Aslanides, J., Firoiu, V., Ewalds, T., Rauh, M., Weidinger, L., Chadwick, M., Thacker, P., et al.: Improving alignment of dialogue agents via targeted human judgements. Preprint (2022). arXiv:2209.14375
80. Liu, H., Sferrazza, C., Abbeel, P.: Chain of hindsight aligns language models with feedback. Preprint (2023). arXiv:2302.02676
81. Guu, K., Lee, K., Tung, Z., Pasupat, P., Chang, M.-W.: Realm: retrieval-augmented language model pre-training. Preprint (2020). arXiv:2002.08909
82. Lewis, P., Perez, E., Piktus, A., Petroni, F., Karpukhin, V., Goyal, N., Küttler, H., Lewis, M., Yih, W.-T., Rocktäschel, T, et al.: Retrieval-augmented generation for knowledge-intensive nlp tasks. Adv. Neural Inf. Process. Syst. (2020)
83. Izacard, G., Grave, É: Leveraging passage retrieval with generative models for open domain question answering. In: Proceedings of the 16th Conference of the European Chapter of the Association for Computational Linguistics (EACL) (2021)
84. Izacard, G., Lewis, P., Lomeli, M., Hosseini, L., Petroni, F., Schick, T., Dwivedi-Yu, J., Joulin, A., Riedel, S., Grave, E.: Few-shot learning with retrieval augmented language models. Preprint (2022). arXiv:2208.03299
85. Borgeaud, S., Mensch, A., Hoffmann, J., Cai, T., Rutherford, E., Millican, K., Van Den Driessche, G.B., Lespiau, J.-B., Damoc, B., Clark, A., et al.: Improving language models by retrieving from trillions of tokens. In: International Conference on Machine Learning, pp. 2206–2240. PMLR (2022)
86. Yu, W., Iter, D., Wang, S., Xu, Y., Ju, M., Sanyal, S., Zhu, C., Zeng, M., Jiang, M.: Generate rather than retrieve: Large language models are strong context generators. International Conference for Learning Representation (ICLR) (2023)

87. Devlin, J., Chang, M.-W., Lee, K., Toutanova, K.: Bert: pre-training of deep bidirectional transformers for language understanding. In: Conference of the North American Chapter of the Association for Computational Linguistics (NAACL) (2019)
88. Lewis, M., Liu, Y., Goyal, N., Ghazvininejad, M., Mohamed, A., Levy, O., Stoyanov, V., Zettlemoyer, L.: Bart: denoising sequence-to-sequence pre-training for natural language generation, translation, and comprehension. In: Annual Meeting of the Association for Computational Linguistics (ACL) (2020)
89. Liu, N.F., Lin, K., Hewitt, J., Paranjape, A., Bevilacqua, M., Petroni, F., Liang, P.: Lost in the middle: how language models use long contexts. Preprint (2023). arXiv:2307.03172

Chapter 5
Augmenting NLP Models with Commonsense Knowledge

Abstract This chapter focuses on augmenting NLP models with commonsense knowledge to enhance their performance in natural language understanding and generation tasks. We begin by discussing the importance of commonsense knowledge in NLP and the challenges faced by NLP models in reasoning with commonsense. We explore different types of commonsense knowledge and reasoning tasks, including multiple-choice tasks, open-ended QA, constrained NLG, and commonsense probing of language models. We then introduce the various techniques for augmenting NLP models with commonsense knowledge. We discuss the use of structured knowledge bases, such as ConceptNet, and the incorporation of graph networks for encoding structured knowledge. We also examine the augmentation of NLP models with un/semi-structured knowledge sources, such as text corpora and the use of dense passage retrieval for open-ended QA. Furthermore, we explore differentiable reasoning methods, such as DrFact, for reasoning with semi-structured knowledge. Finally, we discuss the use of neural knowledge models, such as COMET and LLMs, for incorporating commonsense knowledge. We explore the generation of commonsense knowledge graphs using LLMs and knowledge distillation techniques to create smaller, specialized commonsense models. We also examine the use of large language models for extracting relevant commonsense knowledge for reasoning.

Keywords Commonsense knowledge · Commonsense reasoning · Knowledge graphs · Neural knowledge models

5.1 Commonsense Knowledge and Reasoning for NLP

5.1.1 The Importance of Common Sense in NLP

In the previous chapters, we explored techniques for augmenting NLP models with encyclopedic knowledge sourced from platforms like Wikipedia. However, commonsense knowledge is distinct from encyclopedic knowledge and plays a crucial role in advancing NLP systems and AI models.

While some NLP models, like earlier versions of GPT-3, can understand and reason with questions about factual information, they often struggle with simple commonsense questions. For example, when asked how many wings tigers usually have, GPT-3 incorrectly responds with "two" instead of the correct answer, which is none or zero.

One of the reasons for these mistakes is the abundance of sentences on the web that mention "X has two wings," while there are few explicit statements about commonsense knowledge like "A tiger has no wings." This lack of explicit commonsense knowledge leads to what we call "reporting bias," where people assume that certain obvious information doesn't need to be explicitly stated when communicating.

This reporting bias makes it challenging for data-driven AI systems, such as NLP models, to learn and utilize commonsense knowledge effectively. As a result, NLP models often struggle with language ambiguity and fail to reason with commonsense knowledge, as demonstrated by the Winograd Schema Challenge [1] example where the model fails to correctly identify the referent of the pronoun. For example, "The trophy would not fit in the brown suitcase because it was too big. What is too small?", a human can easily reason and answer that 'it' refers to the 'trophy' instead of the 'suitcase' (Fig. 5.2), but many NLP models cannot do so.

The absence of commonsense knowledge and reasoning abilities can hinder NLP models in various practical use cases, as illustrated in Fig. 5.1. In machine translation, for instance, many sentences can have multiple meanings, and the correct translation depends on the translator's commonsense knowledge. Similarly, in information extraction tasks, entities and relations that are not explicitly stated in the text need to be inferred through commonsense reasoning. Data-driven NLP models heavily rely on the distribution of training data, making it challenging for them to generalize to unseen cases without the aid of commonsense reasoning.

Fig. 5.1 NLP systems need commonsense knowledge and reasoning to handle the ambiguity of natural language

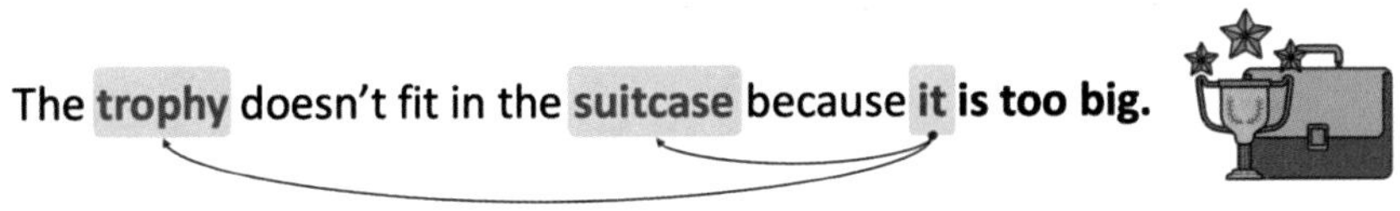

The trophy doesn't fit in the suitcase because it is too big.

Fig. 5.2 An example of coreference resolution that needs common sense to reason about

Furthermore, in vision-language tasks like vision-based question answering, commonsense knowledge is crucial for reasoning with visual context. Without this knowledge, NLP models may struggle to accurately interpret and answer questions based on visual information.

To overcome these challenges, it is essential to develop NLP models that can understand and reason with commonsense knowledge. In this chapter, we will explore different types of commonsense knowledge and reasoning tasks. We will also discuss techniques for augmenting NLP models with commonsense knowledge to enhance their performance in natural language understanding (NLU) and natural language generation (NLG) tasks.

In this section, we first introduce the common sources of commonsense knowledge and then we discuss the reasoning tasks that are used to evaluate the commonsense knowledge of NLP models (Fig. 5.2).

5.1.2 Knowledge Resources

As shown in Fig. 5.3, there are three major types of sources for commonsense knowledge that can be used to improve NLP models in general: (1) structured knowledge, (2) un/semi-structured knowledge, and (3) parametric knowledge.

Primarily, the most widely utilized source of information is structured knowledge. This mainly comprises knowledge graphs such as WikiData (pertaining to general entities about common knowledge), ConceptNet (encompassing commonsense), and Atomics (focusing on event-based knowledge). These sources are meticulously organized and composed of symbolic structures, and most of them are of binary relationship (e.g., relations such as IsA, UsedFor, etc.).

Text corpora are another prevalent source of knowledge. Wikipedia is likely the most informative of these. Wikitionary provides beneficial insight into word comprehension issues, and WikiHow primarily focuses on procedural knowledge. Furthermore, scientific papers contribute significantly to this category. Generic-sKB [2] exemplifies a typical common knowledge corpus.

In addition to these regular knowledge sources, it is necessary to consider a virtual knowledge source: parametric knowledge. This is encoded within the parameters of neural language models. LAMA [3] is a technique used to prompt masked LMs to extract knowledge. COMET [4], a neural knowledge model, was constructed based on ATOMICS [5]. Similarly, many large language models such

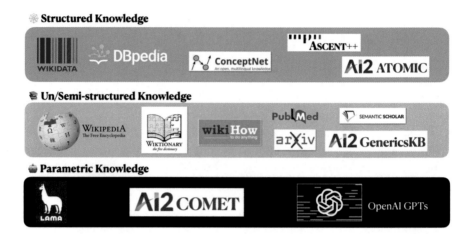

Fig. 5.3 Three primary types of knowledge sources for learning common sense and augment NLP models

as OpenAI GPTs can respond to queries by utilizing their inherent knowledge, such that they can be considered as "virtual" knowledge bases for common sense.

5.1.3 Reasoning Tasks

In this subsection, we present a comprehensive collection of datasets for testing commonsense reasoning ability. They are grouped by different formulations, and cover a wide range of aspects: properties of common objects, real-life situations, elementary science, social skills, etc.

Multiple-Choice Tasks Multiple-choice question answering is a popular task format for commonsense reasoning. Models are given a question and several answer options, and must select the correct choice. Well-known datasets include CommonsenseQA [6], which tests knowledge about everyday objects and events, SocialIQA [7] focusing on social interactions, PhysicalIQA [8] for physical properties and interactions, ARC [9] with science questions for grade-school students, and OpenbookQA [10] modeled after open book exams. Other notable multiple choice datasets are SWAG [11], HellaSWAG [12], WinoGrande [13], X-CSR [14], and RiddleSense [15].

Open-Ended QA Open-ended question answering requires models to output free-form answers to commonsense questions. ProtoQA [16] provides questions about prototypical situations with multiple possible answer categories. OpenCSR [17] focuses on open-ended science questions requiring reasoning over provided background facts.

Constrained NLG Constrained natural language generation tasks evaluate commonsense reasoning by generating text. Given a set of concepts, CommonGen [18] asks models to generate a coherent everyday scenario sentence using those concepts. ComVE SubTask C [19] provides nonsensical statements and requires models to generate explanations of why they are nonsensical. TellMeWhy [20] is similar task that also require NLP models to give reasons behind a scenario. Abductive-NLG [21] tasks a NLG model to generate the missing event between two events, requiring the abductive reasoning ability.

Commonsense probing of LMs Language model probing tasks analyze commonsense knowledge embedded in pre-trained models by predicting masked words. LAMA Probes [3] assess knowledge across different relations. NumerSense [22] specifically targets numerical commonsense across various domains. RICA [23] evaluates robust inference using logical reasoning over commonsense axioms.

5.2 Augmentation with Structured Knowledge

Let's start by discussing structured knowledge and its relevance in natural language processing. One example of structured knowledge is ConceptNet [24], which represents concepts as nodes and their semantic relationships as edges, as shown in Fig. 5.4. However, using knowledge graphs alone to answer natural language questions can be challenging due to difficulties in query understanding and transformation. As a result, when faced with reasoning tasks in the form of multiple-choice question answering, researchers have traditionally relied on models like BERT [25] for learning to reason. However, these language models may struggle with complex reasoning questions because they lack the necessary commonsense knowledge.

This leads us to an important research question: How can we incorporate structured, symbolic knowledge into neural language models?

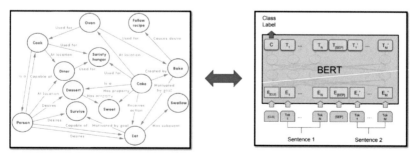

Structured Commonsense Knowledge Graphs   **Pre-Trained Neural Language Models**

Fig. 5.4 Structured knowledge graphs (left) and pre-trained neural language models (right)

5.2.1 Knowledge-Aware Graph Networks Fused with Language Models

5.2.1.1 Overview of the KagNet Framework

KagNet [26] is a typical method that addresses the question of knowledge-aware graph networks. The first step in this approach involves extracting the concepts mentioned in the question and each answer option individually. KagNet then utilizes a knowledge graph, such as ConceptNet, to retrieve a subgraph that contains paths connecting the question concepts and answer concepts (see Fig. 5.5). This subgraph is then encoded using graph networks to obtain a graph vector.

The KagNet is a typical graph neural network design for incorporating structured commonsense knowledge for reasoning with language models. As shown in Fig. 5.6, it consists of three main parts: graph convolutional networks (GCNs) [27] for

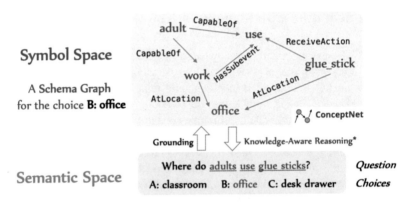

Fig. 5.5 An example of schema graph for commonsense question answering in a multiple-choice setting. The upper part is the symbol space where a subgraph of commonsense KGs represent relevant commonsense knowledge for reasoning about this question

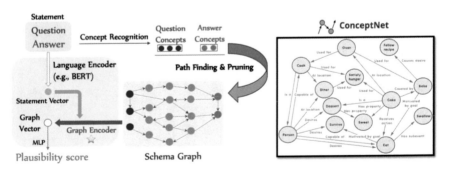

Fig. 5.6 The overall pipeline of KagNet for incorporating external commonsense knowledge into neural language models for reasoning

encoding schema graphs, LSTM-based path encoding for capturing multi-hop relational information, and a hierarchical path-based attention mechanism for modeling relational schema graphs. Simultaneously, the question and answer options are encoded using BERT or other text encoders to obtain a statement vector. The text and graph embeddings are then fused with attention, and a multi-layer perceptron (MLP) is used to learn and predict the plausibility score for each answer option.

This process forms the overall pipeline for KagNet. These knowledge-aware methods are naturally interpretable. Given a question and a prediction, we can examine the attention weights that influence the reasoning process. By identifying the edges and paths with the highest attention weights, we can automatically determine the knowledge used for that prediction. This greatly enhances the interpretability of neural reasoning models.

5.2.1.2 Schema Graph Grounding

Concept Recognition In this step, we match tokens in questions and answers to sets of mentioned concepts (C_q and C_a respectively) from the knowledge graph G. We use `ConceptNet` for its generality. We enhance the naive approach of exact n-gram matching with lemmatization and stop word filtering, and further deal with noise by pruning paths and reducing their importance with attention mechanisms.

Schema Graph Construction ConceptNet is a large set of triples of the form (h, r, t), like (`ice, HasProperty, cold`). We merge the original 42 relation types into 17 types to increase the density of the knowledge graph. A schema graph is a sub-graph g of G that represents the related knowledge for reasoning a given question-answer pair. We construct the graph by finding paths among the mentioned concepts ($C_q \cup C_a$) that are shorter than k concepts.

Path Pruning via KG Embedding To prune irrelevant paths, we use knowledge graph embedding (KGE) techniques, like TransE [28], to pre-train concept embeddings $\mathbf{V}$ and relation type embeddings $\mathbf{R}$. We score a path with the multiplication product of the scores of each triple in the path, and then set a threshold for pruning.

5.2.1.3 KagNet Network Design

Figure 5.7 presents the implementation details of KagNet, which is illustrated step-by-step below.

Graph Convolutional Networks GCNs encode the plain structures of schema graphs by updating node vectors using features from adjacent nodes. This helps to refine concept vectors and capture structural patterns of schema graphs. We initialize the vector for each concept in the schema graph with its pre-trained embedding.

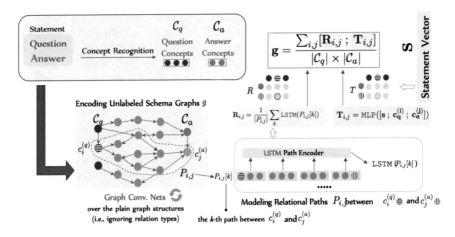

Fig. 5.7 The implementation details of KagNet

Then, we update the vectors at each layer of the GCN using the following equation:

$$h_i^{(l+1)} = \sigma(W_{self}^{(l)} h_i^{(l)} + \sum_{j \in N_i} \frac{1}{|N_i|} W^{(l)} h_j^{(l)})$$

Relational Path Encoding To capture the relational information in schema graphs, they use an LSTM-based path encoder. We represent graphs with respect to the paths between question concepts and answer concepts. Each path is encoded as a sequence of triple vectors using LSTM [29] networks. The resulting latent relation between question and answer concepts is aggregated from all the paths in the schema graph. The vector representation of a schema graph is obtained by aggregating all the path vectors using mean pooling:

$$\mathbf{g} = \frac{\sum_{i,j} [\mathbf{R}_{i,j} \; ; \; \mathbf{T}_{i,j}]}{|C_q| \times |C_a|}$$

Hierarchical Attention Mechanism To selectively aggregate important path vectors and concept-pair vectors, the authors use a hierarchical path-based attention mechanism. This mechanism allows the model to focus on more important paths and concept pairs when constructing the graph representation. The attention scores for the paths and concept pairs are computed using the following equations:

$$\alpha_{(i,j,k)} = \mathbf{T}_{i,j} \, \mathbf{W}_1 \, \mathrm{LSTM}(P_{i,j}[k])$$

$$\hat{\alpha}_{(i,j,\cdot)} = \mathrm{SoftMax}(\alpha_{(i,j,\cdot)})$$

$$\hat{\mathbf{R}}_{i,j} = \sum_k \hat{\alpha}_{(i,j,k)} \cdot \mathrm{LSTM}(P_{i,j}[k])$$

$$\beta_{(i,j)} = \mathbf{s}\,\mathbf{W}_2\,\mathbf{T}_{i,j}$$

$$\hat{\beta}_{(\cdot,\cdot)} = \texttt{SoftMax}(\beta_{(\cdot,\cdot)})$$

$$\hat{\mathbf{g}} = \sum_{i,j} \hat{\beta}_{(i,j)}[\hat{\mathbf{R}}_{i,j}\;;\;\mathbf{T}_{i,j}]$$

In conclusion, the KagNet is a graph neural network module that models relational graphs for relational reasoning using the GCN-LSTM-HPA architecture. It combines the knowledge from the symbolic space of schema graphs and the semantic space of natural language to improve reasoning performance.

5.2.2   Multi-Hop Graph Relational Networks for Knowledge Fusion

KagNet gives a great framework for modeling the relational knowledge of common sense for reasoning with language models. However, KagNet also has some issues, especially for the scalability because there are a very large number of paths from the knowledge graphs, and LSTM-based path encoder is rather inefficient due to the challenges of parallel computing, especially when the length of the paths are too long. This can lower the efficiency a lot due to the fact that LSTM must be executed in a sequential way. Also, the coverage of the retrieved schema graphs can be sub-optimal and miss the useful edges due to heuristics-based pruning.

To address these issues, a follow-up paper, named MHGRN [30], significantly improves the scalability of fusing structured knowledge into LMs by proposing a multi-hop relational graph network to efficiently model the knowledge subgraphs (i.e., schema graphs). The Multi-hop Graph Relation Network (MHGRN) is a novel Graph Neural Network (GNN) architecture that combines the strengths of both GNNs and path-based models. It offers path-level reasoning and interpretability, similar to path-based models, while maintaining the scalability of GNNs. The MHGRN follows the GNN framework, where node features can be initialized with pre-trained weights.

Type-Specific Transformation The model first performs a node type specific linear transformation on the input node features:

$$x_i = U_{\phi(i)} h_i + b_{\phi(i)}, \tag{5.1}$$

where the parameters U and b are specific to the type of node i.

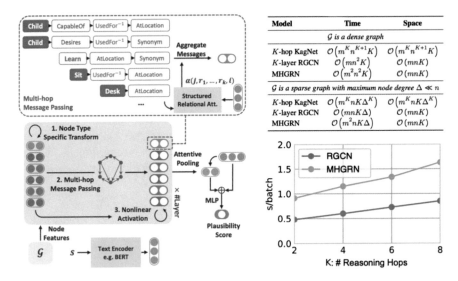

Fig. 5.8 Implementation details of MHGRN and the complexity analysis

Multi-Hop Message Passing As shown in Fig. 5.8, the model passes messages directly over all the relational paths of lengths up to K. The set of valid k-hop relational paths is defined as:

$$\Phi_k = \{(j, r_1, \ldots, r_k, i) \mid (j, r_1, j_1), \cdots, (j_{k-1}, r_k, i) \in \mathcal{E}\} \quad (1 \leq k \leq K). \tag{5.2}$$

The model performs k-hop message passing over these paths, which is a generalization of the single-hop message passing in RGCNs [31]:

$$z_i^k = \sum_{(j, r_1, \ldots, r_k, i) \in \Phi_k} \alpha(j, r_1, \ldots, r_k, i)/d_i^k \cdot W_0^K \cdots W_0^{k+1} W_{r_k}^k \cdots W_{r_1}^1 x_j, \tag{5.3}$$

where $1 \leq k \leq K$ and the $W_r^t (1 \leq t \leq K, 0 \leq r \leq m)$ matrices are learnable, $\alpha(j, r_1, \ldots, r_k, i)$ is an attention score and $d_i^k = \sum_{(j \cdots i) \in \Phi_k} \alpha(j \cdots i)$ is the normalization factor.

Non-linear Activation Finally, shortcut connection and nonlinear activation are applied to obtain the output node embeddings.

$$h_i' = \sigma \left(V h_i + V' z_i \right), \tag{5.4}$$

where V and V' are learnable model parameters, and σ is a non-linear activation function.

Structured Relational Attention The attention score $\alpha(j, r_1, \ldots, r_k, i)$ for all k-hop paths is parameterized without introducing $O(m^k)$ parameters. It is regarded as the probability of a relation sequence $(\phi(j), r_1, \ldots, r_k, \phi(i))$ conditioned on s:

$$\alpha(j, r_1, \ldots, r_k, i) = p\left(\phi(j), r_1, \ldots, r_k, \phi(i) \mid s\right), \tag{5.5}$$

which can be modeled by a probabilistic graphical model, such as conditional random field.

Computation Complexity Analysis The time complexity and space complexity of MHGRN on a sparse graph are both linear with respect to either the maximum path length K or the number of nodes n.

Expressive Power of MHGRN MHGRN is a generalization of RGCN and is capable of directly modeling paths, making it interpretable as are path-based models like RN [32] and KagNet.

Learning, Inference and Path Decoding The learning and inference process of MHGRN for QA tasks aims to determine the plausibility of an answer option $a \in C$ given the question q with the information from both text s and graph $\mathcal{G}$. The model is trained end-to-end jointly with the text encoder (e.g., RoBERTa). During inference, the most plausible answer is predicted by $\text{argmax}_{a \in C} \ \rho(q, a)$. Additionally, a reasoning path can be decoded as evidence for model predictions.

5.2.3 Advanced Graph-Language Fusion Methods

Both KagNet and MHGRN fuse knowledge late when both graph vectors and text embeddings are ready. An advanced method is to fuse the language models and graph encoders in their earlier encoding stages. A more recent model, **QA-GNN** [33], also relies on graph attention network for encoding the graphs. Instead of merging the statement vector at the end of graph encoding, the QA-GNN proposes a special node called context node. It is the purple node Z here. Also, there are two more special relations. The edges between the context node and the question node is of the type c2q and the the edges between the Z and answer nodes are labelled as c2a here. The node Z here represent the semantic features in the text of question and answer, and initialized by a pre-trained LM. This simple modification of the graph structure enables better attention-based reasoning and interpretability. We can clearly see that the reasoning process is iteratively improved over time from the 1st layer to the final layer. And using a BFS to iterate the nodes, we can get very straightforward reasoning chains for explain the model decisions. Similarly, **GreaseLM** [34] focuses on more advanced fusion. They designed special fusion layers to better inject knowledge into each token's representation.

5.2.4 Incorporating Structured Commonsense Knowledge for NLG

Knowledge Graph-Augmented BART (KG-BART) [35] incorporates structured commonsense knowledge from knowledge graphs such as ConceptNet [24] into the pre-trained BART [36] language model for improving commonsense reasoning in natural language generation tasks such as CommonGen [18].

Generative commonsense reasoning requires generating plausible and coherent text describing everyday scenarios from a given set of concepts. However, most existing pre-trained language models like BART often produce implausible or anomalous sentences for this task as they do not consider the relationships between concepts. To address this issue, KG-BART leverages external commonsense knowledge graphs like ConceptNet that provide relational information between concepts. It first constructs a concept-reasoning graph and a hierarchical concept-expanding graph from the knowledge graph to represent dependencies between concepts.

KG-BART follows a standard encoder-decoder architecture but integrates graph representations into the encoding and decoding process through graph attention networks. Specifically, it proposes techniques like subword-to-concept integration and concept-to-subword disintegration to combine token-level text representations with concept-level graph representations. In the encoder, a knowledge graph-augmented transformer enriches token embeddings using graph attention over the concept-reasoning graph. In the decoder, a hierarchical graph attention mechanism propagates intra- and inter-concept relations from the concept-expanding graph for generating more natural text.

Experiments on the CommonGen benchmark for generative commonsense reasoning demonstrate that KG-BART significantly outperforms previous pre-trained language models like BART, GPT-2, and T5 [37]. By incorporating structured commonsense knowledge, KG-BART produces more plausible, coherent, and natural language generations. The paper provides an effective way of integrating external knowledge graphs with pre-trained models for advanced commonsense reasoning in NLG.

5.3 Augmentation with Un/Semi-structured Knowledge

In the previous sections, we have demonstrated the potential of utilizing structured knowledge. However, Knowledge Graphs (KGs) present two significant limitations: their potential incompleteness and their support for only binary relations. For more complex knowledge, we must express them in natural language, which cannot be easily converted into simple triplets. For instance, the knowledge that "Most trees add one new ring for each year of growth" is challenging to represent using binary relations. Therefore, it is crucial to consider unstructured or semi-structured knowledge. These types of corpora provide more natural language descriptions, making them more suitable for complex open-ended questions. In the next sections, we will first introduce the problem setup of open-ended commonsense reasoning and its challenges.

Moreover, using such unstructured knowledge corpora for open-ended reasoning is usually necessary. Unlike multiple-choice question answering which is a closed reasoning setting, open-ended setting requires the NLP models to retrieve multiple pieces of facts for reason.

5.3.1   Open-Ended Reasoning with Commonsense Knowledge

Task Formulation   Figure 5.9 shows an example of open-ended commonsense reasoning. A **corpus** of knowledge facts is denoted as $\mathcal{F}$, and $\mathcal{V}$ is used to denote a vocabulary of **concepts**; both are sets consisting of unique elements. A **fact** $f_i \in \mathcal{F}$ is a sentence that describes generic commonsense knowledge, such as "*trees* remove *carbon dioxide* from the *atmosphere* through *photosynthesis*." A **concept** $c_j \in \mathcal{V}$ is a noun or base noun phrase mentioned frequently in these facts (e.g., 'tree' and 'carbon dioxide'). Concepts are considered identical if their surface forms are the same (after lemmatization).

Given only a **question** q (e.g., "*what can help alleviate global warming?*"), an open-ended commonsense reasoner is supposed to **answer** it by returning a weighted set of concepts, such as $\{(a_1='renewable\ energy', w_1), (a_2='tree', w_2), \ldots\}$, where $w_i \in \mathbb{R}$ is the weight of the predicted concept $a_i \in \mathcal{V}$. To learn interpretable, trustworthy reasoning models, it is expected that models can output intermediate results that justify the reasoning process — i.e., the supporting facts from $\mathcal{F}$. E.g., an **explanation** for 'tree' to be an answer to the question above can be

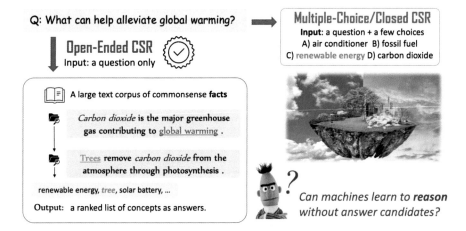

Fig. 5.9   A motivating example for open-ended commonsense reasoning

the combination of two facts: f_1 = "carbon dioxide is the major ..." and f_2 = "trees remove ...".

Implicit Multi-Hop Structures Commonsense questions (i.e., questions that need commonsense knowledge to reason) contrast with better-studied multi-hop factoid QA datasets, e.g., HotpotQA [38], which primarily focus on querying about *evident relations between named entities*. For example, an example multi-hop factoid question can be "which team does the player named 2015 Diamond Head Classic's MVP play for?" Its query structure is relatively clear and *self-evident* from the question itself: in this case the reasoning process can be decomposed into q_1 = "the player named 2015 DHC's MVP" and q_2 = "which team does q_1. answer play for".

The reasoning required to answer commonsense questions is usually more *implicit* and relatively unclear. Consider the previous example, q = 'what can help alleviate global warming?' can be decomposed by q_1 = "what contributes to global warming" and q_2 = "what removes q_1. answer from the atmosphere"—but many other decompositions are also plausible.

In addition, unlike HotpotQA, it is assumed that there are *no ground-truth justifications* for training, which makes OpenCSR even more challenging.

5.3.2 Dense Passage Retrieval for Open-Ended QA

Although off-the-shelf text-search APIs such as ElasticSearch that are based on BM25 sparse features and TF-IDF statistics are applicable for such open-ended question answering setting, their performance is mostly limited due to the lack of semantic understanding and higher-level reasoning skills. In discussing the use of knowledge corpora, such as Wikipedia, one should consider the approach referred to as Dense Passage Retrieval (DPR) [39].

As shown in Fig. 5.10, DPR makes use of two distinct BERT models [25], with one model encoding questions and the other encoding the associated passages. This method allows for pre-computed embedding of all passages via the passage-encoder, which are subsequently saved into a dense index. Given that a question has been put forth, the query-encoder is deployed to ascertain query embeddings, such that "**Maximum Inner Product Search**" (MIPS) can retrieve the passages of greatest relevance to provide additional knowledge for reading and reasoning tasks.

The drawback of DPR resides primarily in its singular-step reasoning restriction, which compromises its capacity for more intricate reasoning tasks. With a view to accommodating more complex questions, it may be deemed requisite to engage at least dual-step processes to ascertain an accurate response. An elementary upscaling of DPR is found in the **Multi-Hop Dense Retrieval** (MDR) method [40], which enables the user to retrieve and reason in an iterative manner until a solution is found.

While the simplicity and user-friendly nature of these dense-only methods are undisputable, they unfortunately overlook the sparse structures embedded within

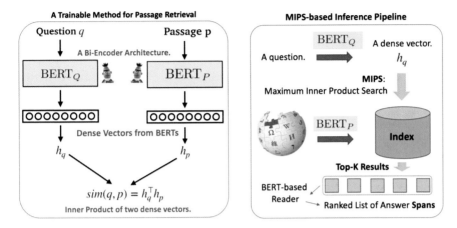

Fig. 5.10 Implementation details of Dense Passage Retrieval (DPR) via Maximum Inner Product Search (MIPS)

the passages. The co-occurrence of entities and concepts acts as a powerful force of sparse features that can efficiently filter out irrelevant retrievals, ensuring that multi-hop reasoning remains accurate. Thus, over-reliance on dense retrieval methodologies alone can result in significant interference, leading to diversion in the reasoning process, ultimately rendering completely irrelevant passages.

5.3.3 Differentiable Reasoning with Semi-Structured Knowledge

We now introduce a method that combines the advantages of dense retrieval and sparse features, named DrFact [17]. We can formulate a knowledge corpus like GenericsKB [2] as facts with the concept mentions. Each fact is a natural language sentence, and each concept is a noun or noun phrase mentioned in a fact. We can consider two facts as "linked" if they contain common concepts and are semantically relevant to each other.

We can pre-compute the dense embedding of facts, similar to DPR [39], and also prepare a sparse matrix to store all the links from concepts to facts, and another sparse matrix to save the fact-to-fact links. The definition of fact links is highly customizable. For instance, one can consider two facts linked if they contain K common keywords. Other conditions such as semantic similarity and entailment scores can also be added. Finally, we define the facts in each step as a super-large sparse vector where non-zero elements save the weights for relevant facts.

During the inference stage, as shown in Fig. 5.11, we can extract the question concepts from the question q and encode its query embeddings. From the dense

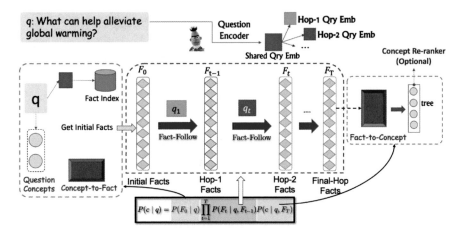

Fig. 5.11 Overview of the DrFact framework for differentiable open-ended commonsense reasoning by multi-hop fact following

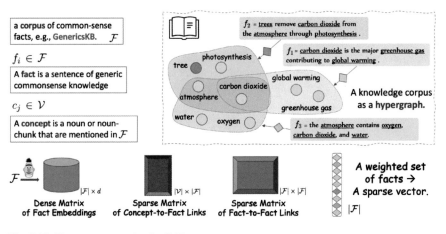

Fig. 5.12 The pre-computation for DrFact

index, we can retrieve the most similar facts, preferably those mentioning question concepts.

More specifically, as illustrated in Fig. 5.12, we can consider the fact-following operation as a dual process. The dense part is similar to DPR, but with a hop-wise query encoder, we can learn to update the query representation over time. The sparse part is based on sparse operations over the matrix that can be differentiably learned and incorporate any heuristics that you may have. In the end, we will mix both results and provide the next-hop facts for reasoning. With a self-following design, one does not need to worry about when to stop reasoning.

Overview *DrFact* [17] is a model for multi-hop reasoning over facts. The model is designed to traverse a hypergraph, where each hyperedge corresponds to a fact in $\mathcal{F}$, connecting the concepts in $\mathcal{V}$ that are mentioned in that fact. The reasoning process is modeled as a probabilistic process over T hops, represented as

$$P(c \mid q) = P(c \mid q, F_T) \prod_{t=1}^{T} P(F_t \mid q, F_{t-1}) P(F_0 \mid q).$$

Pre-computed Indices *DrFact* uses pre-computed indices (see Fig. 5.12), including a Dense Neural Fact Index D, a Sparse Fact-to-Fact Index S, and a Sparse Index of Concept-to-Fact Links E. The Dense Neural Fact Index D is pre-trained using a bi-encoder architecture over BERT, which learns to maximize the score of facts that contain correct answers to a given question. The Sparse Fact-to-Fact Index S pre-computes the sparse links between facts by a set of connection rules. The Sparse Index of Concept-to-Fact Links E encodes the co-occurrences between each fact and its mentioned concepts into a sparse matrix.

Differentiable Fact-Following Operation As shown in Fig. 5.13, the model introduces a differentiable fact-following operation, $P(F_t \mid F_{t-1}, q)$, which combines both neural embeddings of the facts and their symbolic connections in the hypergraph. This operation involves two parallel sub-steps: sparse retrieval and dense retrieval. The sparse retrieval uses a fact-to-fact sparse matrix to obtain possible next-hop facts, while the dense retrieval uses a maximum inner product search (MIPS) over the dense fact embedding index D. The results from both retrievals are combined using element-wise multiplication.

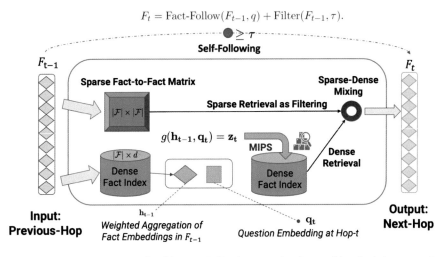

Fig. 5.13 Implementation details of the Fact-Following operation that combines both dense search and sparse search, as well as the integration of self-following ability

Self-Following and Initial Facts The authors also introduce a concept of "self-following", which means that F_t contains highly-relevant facts for all distances $t' < t$. This improves the model's performance when there are variable numbers of "hops" for different questions. The set of initial facts F_0 is computed differently, as they are produced using the input question q.

Auxiliary Learning with Distant Evidence *DrFact* also uses auxiliary learning with distant evidence to guide multi-hop reasoning models during training. The authors use dense retrieval based on the training questions to get some noisy yet still helpful supporting facts. The final learning objective is to optimize the sum of the cross-entropy loss l between the final weighed set of concepts A and the answer set A^*, as well as the auxiliary loss from distant evidence, represented as

$$\mathcal{L} = l(A, A^*) + \frac{1}{T}\sum_{t=1}^{T} l(F_t, F_t^*).$$

This approach allows the model to collect the set of supporting facts at each hop position, denoted as $\{F_1^*, F_2^*, \ldots, F_T^*\}$.

Summary In summary (Fig. 5.14), we have discussed several methods for knowledge augmentation in NLP. BM25 is a simple, unsupervised sparse method for retrieving facts. DPR and MDP are dense-only methods and do not consider the structures in the corpus. DrKIT [41] is a method similar to DrFact, but it focuses on entity-mention embeddings, which is less suitable in corpora like GenericsKB. DrFact is a more flexible multi-hop reasoning method and is very suitable for sentence-based knowledge corpora.

Methods	BM25	DPR/ MDR	DrKIT	DrFact
Knowledge Structure	A set of documents	A set of documents	Mention-Entity Bipartite Graph	Concept-Fact Hypergraph
Multi-hop Reasoning Formulation	-	- / Multiple-Round	Entity-Following	Fact-Following
Index for Dense Retrieval	-	Passage Embeddings	Mention Embedding	Fact Embeddings
Sparse Retrieval Method	TF-IDF based Index+ BM25 Ranking Func.	-	Entity-Mention Cooccurrence	Fact-to-Fact Matrix
Multi-Hop Questions	-	- / Single model	Aggregating Multiple Models	A single model w/ Self-Following
Intermediate Supervision	-	-	N/A	Distant Supervision

Fig. 5.14 An overall comparison of different retrieval methods for open-ended reasoning

5.4   Augmentation with Neural Knowledge Models

In the previous sections, we have primarily discussed textual knowledge sources. These sources are generally accurate and easily modifiable, making them trustworthy and verifiable. However, they can also be incomplete and challenging to query, necessitating the development of numerous methods for incorporating them into neural models over the years.

Can neural networks be used as knowledge sources for common sense and improve NLP models? We first show that one can learn to train language models as knowledge models. The advent of models like ChatGPT and GPT-4 suggests that large language models (LLMs) can serve as promising knowledge sources. They are highly generalizable, cover a wide range of domains, and can be queried easily using natural language. However, they can also generate hallucinations, leading to unusual errors. In this section, we will introduce methods to better utilize LLMs as knowledge sources for reasoning.

5.4.1   Neural Knowledge Models of Common Sense

COMET [4] is a typical generative model for automatically constructing commonsense knowledge graphs. Commonsense knowledge, such as knowing that "going to the store requires having money", is important for improving natural language processing systems but difficult to represent.

COMET is created by fine-tuning a pre-trained transformer language model on an existing commonsense knowledge graph to generate novel knowledge tuples in the form (subject, relation, object), such as (personX goes to store, xIntent, to buy food). The key idea is to leverage the implicit commonsense captured in large language models through pre-training on text corpora. COMET learns to adapt these representations to generate explicit commonsense knowledge.

Experiments on two commonsense KGs, ATOMIC and ConceptNet, demonstrate COMET's ability to produce novel, high-quality knowledge tuples. Human evaluation found up to 77.5% precision on ATOMIC and 91.7% on ConceptNet. COMET also generates more diverse commonsense compared to prior extractive methods. The authors conclude that generative commonsense models like COMET are a promising approach for automatic knowledge base construction. By leveraging pre-trained language models, COMET takes a step toward enabling deeper natural language understanding through common sense reasoning. A follow-up paper further extend the COMET with more high-quality data and thus create ATOMIC-COMET [42] that is even more powerful as knowledge models of common sense.

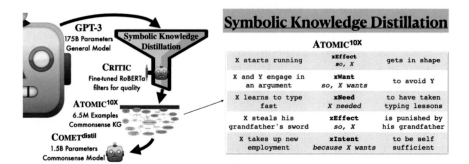

Fig. 5.15 Symbolic knowledge distillation (SKD) aims to distill a local small knowledge model by training language models with symbolic data generated by LLMs such as GPT-3 and beyond

5.4.2 LLMs as Knowledge Models

Symbolic Knowledge Distillation from LLMs LLMs, such as OpenAI's GPT models, can be costly to implement in production, and their internal weights and data are not transparent, which is detrimental to open and reproducible research. A natural question arises: can we distill knowledge from these models, save the intermediate data in text form, and train a smaller, more manageable knowledge model to mimic the larger GPT? The method of Symbolic Knowledge Distillation (SKD) [43], proposed by AI2, suggests that this is not only possible but also promising.

SKD is a typical framework for distilling commonsense knowledge from large language models (LLMs) like GPT-3 into specialized commonsense models (see Fig. 5.15). The key idea is to leverage the implicit commonsense captured in LLMs' pre-training, but focus it into an explicit commonsense model. This is done by prompting the LLM to generate a large commonsense knowledge graph, which is then used to train a smaller model to specialize in commonsense reasoning. A key component is the use of a critic model to filter the LLM's generations, improving the quality of the resulting knowledge graph. With careful prompt engineering, the authors show SKD can produce commonsense graphs exceeding human-authored resources in scale, accuracy and diversity.

The resulting commonsense model, COMET-distill, surpasses prior models trained on human-written knowledge, despite being 100x smaller than the LLM teacher. This demonstrates the promise of SKD for extracting and focusing the diffuse commonsense abilities of large pre-trained models.

By collaboratively using LLMs and humans, where the former as commonsense generators, and the latter as evaluators for the critic, SKD provides a scalable path to high-quality commonsense knowledge. The authors conclude that LLMs and limited human judgment can outperform purely human-authored resources. The success of SKD indicates that LLMs and humans can collaborate effectively—with machines generating candidates and humans evaluating—to create commonsense

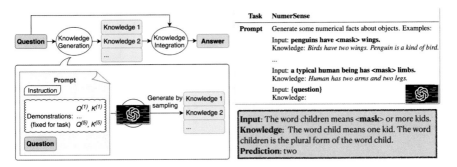

Fig. 5.16 Generated knowledge prompting is a method to use LLMs such as GPT-3 as a commonsense knowledge source for reasoning

knowledge graphs. The method provides a scalable path to expanding resources for commonsense AI systems.

Generated Knowledge Prompting for LLMs Another approach is to use GPTs as a knowledge source to achieve optimal performance. The most straightforward way to do this is to create knowledge-querying prompts to extract relevant knowledge for a question, and then use a smaller local model for reasoning. This approach, known as Generated Knowledge Prompting (GKP) [44], as depicted in Fig. 5.16, works as follows: given a question like "The word children means [mask] or more kids.", we first use few-shot in-context demonstrations to prompt GPT to provide the knowledge—"The word child means one kid, and the word children is its plural form." This knowledge is then shown to a reasoning model, which can then better answer the question. This is a simple way to query knowledge from GPTs and use the retrieved knowledge as additional context for local reasoning with smaller models like T5 [37].

While GKP is powerful, it can often generate noisy knowledge from GPTs, known as hallucination issues. For instance, when asked the question, "How many legs does an easel have?", GPT might incorrectly associate a human with an easel, leading the reasoning model to incorrectly answer that an easel has two legs instead of three. To address this issue, the Rainier [45] method proposes using reinforcement learning to train the local knowledge reasoning module to accept only correct and relevant knowledge and reject hallucination responses.

5.5 Conclusion

In this chapter, we explored different techniques for augmenting NLP models with commonsense knowledge to enhance their language understanding and reasoning abilities. We discussed the importance of commonsense knowledge in NLP and the challenges faced by models in reasoning with commonsense. We examined different sources of commonsense knowledge, including structured knowledge graphs, un/semi-structured knowledge corpora, and neural knowledge models.

5.5.1 Summary

Structured Knowledge Graphs We discussed the use of structured knowledge graphs, such as ConceptNet, as a source of commonsense knowledge. We explored methods like KagNet and MHGRN that use graph neural networks to encode the structured knowledge and reason with it. These models have shown promising results in enhancing NLP models' performance in commonsense reasoning tasks.

Un/Semi-Structured Knowledge Corpora We also explored the use of un/semi-structured knowledge corpora for open-ended reasoning. We discussed the Dense Passage Retrieval (DPR) method for retrieving relevant passages from knowledge corpora and the DrFact method for performing differentiable reasoning over the knowledge graph to answer open-ended questions. These methods leverage the strengths of both dense and sparse representations to improve reasoning performance.

Neural Knowledge Models Furthermore, we discussed the use of neural knowledge models like COMET and Symbolic Knowledge Distillation (SKD) to leverage large language models as knowledge sources. COMET generates commonsense knowledge graphs from pre-trained language models, while SKD distills the knowledge from large language models into smaller, more specialized commonsense models. These methods provide scalable ways to create and utilize commonsense knowledge from neural models.

5.5.2 Future Directions

There are still many interesting future research directions in this area. Here are some open questions to consider:

Integration of Different Knowledge Sources How can we effectively combine and merge different sources of knowledge, including structured knowledge graphs, un/semi-structured corpora, and neural knowledge models, into a unified framework that leverages the strengths of each source?

Faithfulness and Interpretability How can we design reasoning models that are not only accurate but also faithful to the underlying knowledge sources? How can we ensure that the reasoning process is transparent and interpretable to users and stakeholders? Can we develop methods to provide explanations or justifications for the model's reasoning?

Real-World Situations and Social Interactions How can we collect and encode knowledge about real-world situations, social interactions, and nuanced human behavior into NLP models? Can we develop methods to reason with this type of knowledge for better understanding and generation of natural language?

Transfer and Generalization How can we improve the transfer and generalization capabilities of NLP models when reasoning with commonsense knowledge? Can we develop methods to enable models to reason with commonsense across different domains and adapt to new contexts?

Evaluation and Benchmarking How can we develop robust evaluation metrics and benchmarks for assessing the performance of commonsense reasoning models? How can we create comprehensive datasets that cover a wide range of commonsense reasoning tasks for training and evaluation purposes?

Addressing these open questions will require interdisciplinary research efforts combining natural language processing, knowledge representation, cognitive science, and machine learning. The answers to these questions will contribute to the development of more intelligent and reliable NLP systems that can reason with commonsense knowledge effectively.

In conclusion, augmenting NLP models with commonsense knowledge is crucial for improving their language understanding and reasoning abilities. The techniques discussed in this chapter provide valuable insights and directions for incorporating commonsense knowledge into NLP models and advancing the field of natural language processing.

References

1. Levesque, H.J., Davis, E., Morgenstern, L.: The winograd schema challenge. In: AAAI Spring Symposium: Logical Formalizations of Commonsense Reasoning (2011)
2. Bhakthavatsalam, S., Anastasiades, C., Clark, P.: Genericskb: a knowledge base of generic statements (2020). ArXiv, abs/2005.00660
3. Petroni, F., Rocktäschel, T., Lewis, P., Bakhtin, A., Wu, Y., Miller, A.H., Riedel, S.: Language models as knowledge bases? (2019) ArXiv, abs/1909.01066
4. Bosselut, A., Rashkin, H., Sap, M., Malaviya, C., Celikyilmaz, A., Choi, Y.: Comet: commonsense transformers for automatic knowledge graph construction. In Annual Meeting of the Association for Computational Linguistics (2019)
5. Sap, M., Le Bras, R., Allaway, E., Bhagavatula, C., Lourie, N., Rashkin, H., Roof, B., Smith, N.A., Choi, Y.: Atomic: an atlas of machine commonsense for if-then reasoning. In: Proceedings of the AAAI Conference on Artificial Intelligence, vol. 33, pp. 3027–3035 (2019)
6. Talmor, A., Herzig, J., Lourie, N., Berant, J.: Commonsenseqa: a question answering challenge targeting commonsense knowledge (20190. ArXiv, abs/1811.00937
7. Sap, M., Rashkin, H., Chen, D., Le Bras, R., Choi, Y.: Social IQa: commonsense reasoning about social interactions. In: Proceedings of the 2019 Conference on Empirical Methods in Natural Language Processing and the 9th International Joint Conference on Natural Language Processing (EMNLP-IJCNLP), Hong Kong, China, November 2019, pp. 4463–4473. Association for Computational Linguistics (2019)

8. Bisk, Y., Zellers, R., Le Bras, R., Gao, J., Choi, Y.: Piqa: reasoning about physical commonsense in natural language (2019). ArXiv, abs/1911.11641

9. Clark, P., Cowhey, I., Etzioni, O., Khot, T., Sabharwal, A., Schoenick, C., Tafjord, O.: Think you have solved question answering? try arc, the ai2 reasoning challenge (2018). ArXiv, abs/1803.05457

10. Mihaylov, T., Clark, P., Khot, T., Sabharwal, A.: Can a suit of armor conduct electricity? a new dataset for open book question answering. In: Conference on Empirical Methods in Natural Language Processing (2018)

11. Zellers, R., Bisk, Y., Schwartz, R., Choi, Y.: SWAG: a large-scale adversarial dataset for grounded commonsense inference. In Proceedings of the 2018 Conference on Empirical Methods in Natural Language Processing, Brussels, Belgium, October–November 2018, pp. 93–104. Association for Computational Linguistics (2018)

12. Zellers, R., Holtzman, A., Bisk, Y., Farhadi, A., Choi, Y.: HellaSwag: can a machine really finish your sentence? In: Proceedings of the 57th Annual Meeting of the Association for Computational Linguistics, Florence, Italy, July 2019, pp. 4791–4800. Association for Computational Linguistics (2019)

13. Sakaguchi, K., Le Bras, R., Bhagavatula, C., Choi, Y.: Winogrande: an adversarial winograd schema challenge at scale (2019). ArXiv, abs/1907.10641

14. Lin, B.Y., Lee, S., Qiao, X., Ren, X.: Common sense beyond English: evaluating and improving multilingual language models for commonsense reasoning (2021). ArXiv, abs/2106.06937

15. Lin, B.Y., Wu, Z., Yang, Y., Lee, D.-H., Ren, X.: Riddlesense: reasoning about riddle questions featuring linguistic creativity and commonsense knowledge. In: Findings (2021)

16. Boratko, M., Li, X.L., Das, R., O'Gorman, T.J., Le, D., McCallum, A.: newblock Protoqa: a question answering dataset for prototypical common-sense reasoning (2020). ArXiv, abs/2005.00771

17. Lin, B.Y., Sun, H., Dhingra, B., Zaheer, M., Ren, X., Cohen, W.W.: Differentiable open-ended commonsense reasoning (2020). ArXiv, abs/2010.14439

18. Lin, B.Y., Zhou, W., Shen, M., Zhou, P., Bhagavatula, C., Choi, Y., Ren, X.: CommonGen: a constrained text generation challenge for generative commonsense reasoning. In: Findings of the Association for Computational Linguistics: EMNLP 2020 (2020)

19. Wang, C., Liang, S., Jin, Y., Wang, Y., Zhu, X., Zhang, Y.: Semeval-2020 task 4: commonsense validation and explanation. In: International Workshop on Semantic Evaluation (2020)

20. Lal, Y.K., Chambers, N., Mooney, R., Balasubramanian, N.: TellMeWhy: a dataset for answering why-questions in narratives. In: Findings of the Association for Computational Linguistics: ACL-IJCNLP 2021, Online, August 2021, pp. 596–610. Association for Computational Linguistics (2021)

21. Bhagavatula, C., Le Bras, R., Malaviya, C., Sakaguchi, K., Holtzman, A., Rashkin, H., Downey, D., Yih, S.W.T., Choi, Y.: Abductive commonsense reasoning (2019). ArXiv, abs/1908.05739

22. Lin, B.Y., Lee, S., Khanna, R., Ren, X.: Birds have four legs?! numersense: probing numerical commonsense knowledge of pre-trained language models. In: Conference on Empirical Methods in Natural Language Processing (2020)

23. Zhou, P., Khanna, R., Lin, B.Y., Ho, D., Pujara, J., Ren, X.: Rica: Evaluating robust inference capabilities based on commonsense axioms. In: Conference on Empirical Methods in Natural Language Processing (2020)

24. Speer, R., Chin, J., Havasi, C.: Conceptnet 5.5: an open multilingual graph of general knowledge. In Thirty-First AAAI Conference on Artificial Intelligence (2017)

25. Devlin, J., Chang, M.-W., Lee, K., Toutanova, K.: Bert: Pre-training of deep bidirectional transformers for language understanding. In: NAACL-HLT (2019)

26. Lin, B.Y., Chen, X., Chen, J., Ren, X.: KagNet: knowledge-aware graph networks for commonsense reasoning. In: Proceedings of the 2019 Conference on Empirical Methods in Natural Language Processing and the 9th International Joint Conference on Natural Language Processing (EMNLP-IJCNLP), Hong Kong, China, November 2019, pp. 2829–2839. Association for Computational Linguistics (2019)

27. Kipf, T., Welling, M.: Semi-supervised classification with graph convolutional networks (2016). ArXiv, abs/1609.02907
28. Bordes, A., Usunier, N., García-Durán, A., Weston, J., Yakhnenko, O.: Translating embeddings for modeling multi-relational data. In Neural Information Processing Systems (2013)
29. Hochreiter, S., Schmidhuber, J.: Long short-term memory. Neural Comput. **9**, 1735–1780 (1997)
30. Feng, Y., Chen, X., Lin, B.Y., Wang, P., Yan, J., Ren, X.: Scalable multi-hop relational reasoning for knowledge-aware question answering (2020). ArXiv, abs/2005.00646
31. Schlichtkrull, M., Kipf, T., Bloem, P., van den Berg, R., Titov, I., Welling, M.: Modeling relational data with graph convolutional networks. In: Extended Semantic Web Conference (2017)
32. Santoro, A., Raposo, D., Barrett, D.G.T., Malinowski, M., Pascanu, R., Battaglia, P.W., Lillicrap, T.P.: A simple neural network module for relational reasoning. In: Neural Information Processing Systems (2017)
33. Yasunaga, M., Ren, H., Bosselut, A., Liang, P., Leskovec, J.: Qa-gnn: reasoning with language models and knowledge graphs for question answering. In: North American Chapter of the Association for Computational Linguistics (2021)
34. Zhang, X., Bosselut, A., Yasunaga, M., Ren, H., Liang, P., Manning, C.D., Leskovec, J.: Greaselm: graph reasoning enhanced language models for question answering (2022). ArXiv, abs/2201.08860
35. Liu, Y., Wan, Y., He, L., Peng, H., Yu, P.S.: Kg-bart: Knowledge graph-augmented bart for generative commonsense reasoning. In: AAAI Conference on Artificial Intelligence (2020)
36. Lewis, M., Liu, Y., Goyal, N., Ghazvininejad, M., Mohamed, A., Levy, O., Stoyanov, V., Zettlemoyer, L.: Bart: denoising sequence-to-sequence pre-training for natural language generation, translation, and comprehension (2019). ArXiv, abs/1910.13461
37. Raffel, C., Shazeer, N., Roberts, A., Lee, K., Narang, S., Matena, M., Zhou, Y., Li, W., Liu, P.J.: Exploring the limits of transfer learning with a unified text-to-text transformer. J. Mach. Learn. Res. **21**(140), 1–67 (2020)
38. Yang, Z., Qi, P., Zhang, S., Bengio, Y., Cohen, W.W., Salakhutdinov, R., Manning, C.D.: HotpotQA: a dataset for diverse, explainable multi-hop question answering. In EMNLP (2018)
39. Karpukhin, V., Oguz, B., Min, S., Lewis, P., Wu, L.Y., Edunov, S., Chen, D., Yih, W.T.: Dense passage retrieval for open-domain question answering (2020). ArXiv, abs/2004.04906
40. Xiong, W., Li, X.L., Iyer, S., Du, J., Lewis, P.., Wang, W.Y., Mehdad, Y., Yih, W.T., Riedel, S., Kiela, D., Oguz, B.: Answering complex open-domain questions with multi-hop dense retrieval (2020). ArXiv, abs/2009.12756
41. Dhingra, B., Zaheer, M., Balachandran, V., Neubig, G., Salakhutdinov, R., Cohen, W.W.: Differentiable reasoning over a virtual knowledge base (2020). ArXiv, abs/2002.10640
42. Hwang, J.D., Bhagavatula, C., Le Bras, R., Da, J., Sakaguchi, K., Bosselut, A., Choi, Y.: Comet-atomic 2020: on symbolic and neural commonsense knowledge graphs. In: AAAI Conference on Artificial Intelligence (2020)
43. West, P., Bhagavatula, C., Hessel, J., Hwang, J.D., Jiang, L., Le Bras, R., Lu, X., Welleck, S., Choi, Y.: Symbolic knowledge distillation: from general language models to commonsense models. In: North American Chapter of the Association for Computational Linguistics (2021)
44. Liu, J., Liu, A., Lu, X., Welleck, S., West, P., Le Bras, R., Choi, Y., Hajishirzi, H.: Generated knowledge prompting for commonsense reasoning. In: Annual Meeting of the Association for Computational Linguistics (2021)
45. Liu, J., Hallinan, S., Lu, X., He, P., Welleck, S., Hajishirzi, H., Choi, Y.: Rainier: reinforced knowledge introspector for commonsense question answering (2022). ArXiv, abs/2210.03078

Chapter 6
Summary and Future Directions

Abstract This chapter summarizes the breakthrough of knowledge-augmented NLP techniques in three lines: natural language understanding, natural language generation, and commonsense reasoning. It discusses a few research challenges such as the heterogeneity of knowledge, global-vs-local knowledge, and scaling of knowledge augmentation, as well as future directions such as knowledge augmentation for structured data tasks (e.g., information extraction), faithfulness, and diverse generation.

Keywords Recent breakthrough · Research challenges · Future directions · Knowledge augmentation

6.1 Summary

This book discusses machine learning techniques for NLP tasks. The techniques are described as learning a function that best maps input variables (X) to an output variable (Y). However, the dependency of Y on X may not be fully direct— intermediate variables are needed in the learning process. The intermediate variables are so called "knowledge" $K(X)$, or more briefly, K. Knowledge-augmented NLP assumes that learning the probability $p(Y|X, K)$ would be easier and more generalizable than learning $p(Y|X)$.

Knowledge bridges the gap between input and output. The gaps can be seen in many NLP tasks. For example, sentiment analysis aims at the gap between a product review sentence and a user sentiment category; machine translation fills the gap between a source-language sentence and a target-language sentence; fact verification is to find the connection between a factual statement and an explanation to true or false; and question answering is naturally bridges the gap between a question and an answer that may not have high semantic similarity.

Where can we find "knowledge" for NLP tasks? Lots of data repositories have been used for different task domains, such as (1) OMCS and ConceptNet for commonsense reasoning, (2) Wikipedia, WikiData, and Wiktionary for encyclopedia, (3) Freebase, DBpedia, and YAGO for general domain, and (4) UMLS, ArnetMiner, and DBLP for scientific domains. Moreover, large language models such as GPT-3 and ChatGPT can also be used to generate knowledge data.

In this book, we focus on three categories of NLP tasks: (1) natural language understanding (NLU), (2) natural language generation (NLG), and (3) commonsense reasoning. We discuss a few questions about knowledge for each task category: (1) What are the specific tasks that need knowledge? (2) Where does the knowledge come from? (3) How is knowledge augmented to language models for the tasks?

6.1.1   Knowledge-augmented NLU

Understanding natural language requires knowledge about entities and concepts, especially for information extraction (IE) and question answering (QA). There are many subtypes of such tasks: (1) named entity recognition (NER), entity linking, slot filling, relation extraction, and fact verification for IE, and (2) Open-domain QA, Commonsense QA, and Knowledge-base QA. The knowledge augmentation methods are entity linking based methods (e.g., ERNIE, KEAR, EaE, FILM, K-BERT) and retrieval-based methods (e.g., DPR, REALM, REINA, RETRO, WebGPT).

6.1.2   Knowledge-augmented NLG

Text generation is widely used in question answering, dialog system, reasoning, machine translation, image captioning, sports news creation. Specifically, the techniques are used to generate questions, answers, responses, explanations, summaries, translations, paraphrases, image captions, news articles, etc. The methods are knowledge graph-based methods (e.g., GRF, CCM, MoKGE) and grounded document-based methods (e.g., RAG, RE-T5, CMR).

6.1.3   Knowledge-augmented Commonsense Reasoning

Commonsense reasoning is a human-like ability to make presumptions about the type and essence of ordinary situations humans encounter every day. OMCS and ConceptNet are often used to enhance language models for the reasoning tasks. The methods include KagNet, MHGRN, QA-GNN, GreaseLM, GSC, CommonGen, KFCNet, KG-BART, I&V, and DrFact.

6.2 Challenges and Future Opportunities

Two open problems in this research field are (1) effectiveness and efficiency of knowledge augmentation methods and (2) benefits to language techniques from structured knowledge. Discussions on three concrete challenges are offered below for each problem.

6.2.1 Augmenting with Heterogeneous Knowledge

Any type of data that bridges the gap between input and output can be considered and used as "knowledge" in NLP techniques. There are several data types of knowledge sources. The most common type is text. Given an input query, a text retriever finds sentences that are related to the input and lead language models to predict the output.

Besides the unstructured knowledge, there are many types of structured or semistructured knowledge. For example, relational databases provide a rich set of attributes in data table columns about relevant data objects. Fact triplets describe the factual relationships between entities and concepts. Knowledge graphs contain a large number of entity nodes and their relational links. Neighbors, paths, and subgraphs can be fetched to augment the input of language models. Taxonomies are a special type of knowledge graphs that describe the parent-child relationship between entities or concepts in terms of types or categories. Ontologies are also commonly used. Dictionaries have key-value pairs about the descriptions or definitions of rare words or concepts. Finally, we have multimedia data such as images, audio, and videos that can be used as knowledge sources. Multimedia data provides knowledge in different views from the text form.

Different relevance scoring functions, search algorithms, and retrieval techniques have been developed for different types of data. Given a query, all of them aim at finding a (small) subset of information relevant to the query, and the information can be transformed to an embedding vector and integrated into a deep learning architecture of language models. However, each would require an exhaustive process to search the optimal settings or hyperparameters. And there is little study that investigates the synergy among the heterogeneous knowledge sources. It is unknown how the multiple types of knowledge data can be fully or maximally used to augment the language models.

Considering the complexity of retrieving and integrating heterogeneous knowledge, a unified approach will be very useful. Suppose all the types of knowledge can be transformed into one type, e.g., text. Then a simple retrieval augmentation approach will be able to utilize the information originally from all the knowledge sources. For example, a large-scale commonsense corpus can be created from commonsense corpora, commonsense knowledge graphs, commonsense annotated datasets, etc. via data-to-text templates or models.

6.2.2 Augmenting with Knowledge Beyond Passages

Structured knowledge might not naturally exist but was constructed via information extraction (IE) methods. Given a query, the methods can locate it to a specific sentence in a large-scale corpus which is then turned into structured data to augment language models. However, the augmentation is limited to the local context, and global knowledge is often ignored. The global context can address multiple issues of NLP techniques, such as contradictions in ideas, singular-plural disagreement, point-of-view inconsistency, and inconsistency across discourses, tables, and graphs. A potential idea is to employ document-level IE or corpus-level IE methods to capture long-term dependencies.

6.2.3 Scaling Knowledge Augmentation

Scaling up knowledge coverage would require incorporating dynamic facts or knowledge, improving model efficiency, and addressing noise from external sources. First, external knowledge should be learned in a continuous way, known as lifelong learning. The dynamic facts must be extracted and used, e.g., Twitter's CEO from Dorsey, to Musk, and to Yaccarino. Second, we desire multiple knowledge pieces that are complementary to each other and interconnected around the gap between input and output. Third, sometimes a large number of passages are needed—when retrieval augmentation methods are not efficient, the structured knowledge such as knowledge graphs may filter out less related passages. Fourth, indexing and searching in massive text is computationally expensive, when we extend the knowledge source from Wikipedia to web-scale corpora. Lastly, noise exists due to imperfect automatic IE and knowledge acquisition methods. For example, the F-score is only around 75% on end-to-end IE in many domains. The reliability of information sources varies at different levels. And it is often hard to find highly reliable sources for specific domains such as biomedicine and chemistry.

6.2.4 Knowledge Augmentation for Structured Data Tasks

Information extraction tasks usually aim to turn unstructured data into some structures. There are still a few types of information extraction tasks that aim to complete or expand the structured data using the unstructured data. For example, given a large-scale text corpus and a small seed taxonomy, the task is to create a large taxonomy; however, taxonomy construction models may limit to the concept pool identified from the text corpus. A great number of concepts on the taxonomy may not appear in the corpus but all of their words may be found frequently in the corpus. So it is important to learn from the structured knowledge (e.g., seed taxonomy or intermediate taxonomy) and fuse the structured and unstructured knowledge for the task. One crazy idea is to generate concepts word by word on the taxonomy, which utilizes all the information from the input data.

6.2.5 Knowledge Augmentation for Faithfulness

Language models suffer from hallucination, especially in natural language generation (NLG) tasks. For example, a notable portion of abstractive summaries contain unfaithful information by state of the art models. So it will be useful to improve the factual correctness of the summaries with knowledge graphs and/or knowledge bases. Moreover, it is important to filter out popularity bias and incorporate more scenario-specific and domain-specific knowledge.

6.2.6 Knowledge Augmentation for NLG Diversity

NLG models usually rely on beam search and/or nucleus sampling to create different kinds of outputs. However, such techniques cannot essentially create diverse outputs. Content diversity requires different perspectives and high accuracy. One idea is to leverage one-to-many-words dictionaries and one-to-many-neighbors KG. Different relevant passages, neighbors, paths, and subgraphs suggest different ideas to create different generated content. For example, random walks on commonsense knowledge graphs may inspire language models to generate different explanation sentences for the same counterfactual statement.

Printed in the United States
by Baker & Taylor Publisher Services